A2 UNIT 1

STUDENT GUIDE

CCEA

Geography

Physical processes, landforms and management

Tim Manson

Alistair Hamill

HODDER
EDUCATION
AN HACHETTE UK COMPANY

Orders: please contact Hachette UK Distribution, Hely Hutchinson Centre, Milton Road, Didcot, Oxfordshire, OX11 7HH. Telephone: +44 (0)1235 827827. Email: education@hachette.co.uk. Lines are open from 9 a.m. to 5 p.m., Monday to Friday. You can also order through our website: www.hoddereducation.co.uk.

ISBN 978-1-4718-6312-7

First published in 2017 by
Hodder Education,
An Hachette UK Company
Carmelite House
50 Victoria Embankment
London EC4Y 0DZ

www.hoddereducation.co.uk

First printed 2017

Impression number 6

Year 2023

Cover photo: Alistair Hamill

Typeset by Integra Software Services Pvt Ltd, Pondicherry, India

Printed and bound by CPI Group (UK) Ltd, Croydon, CR0 4YY

Hachette UK's policy is to use papers that are natural, renewable and recyclable products and made from wood grown in well-managed forests and other controlled sources. The logging and manufacturing processes are expected to conform to the environmental regulations of the country of origin.

Contents

Getting the most from this book 4

About this book 5

Content Guidance

Option A Plate tectonics: theory and outcomes 6

    Plate tectonics: margins and landforms 6

    Volcanic activity and its management 15

    Seismic activity and its management 22

Option B Tropical ecosystems: nature and sustainability 28

    Locations and climates of major tropical biomes 28

    Management and sustainability in arid/semi-arid tropical ecosystems 37

    Management and sustainability in the tropical forest environment 41

Option C Dynamic coastal environments 49

    Coastal processes and features 49

    Regional coastlines 60

    Coastal management and sustainability 64

Option D Climate change: past and present 71

    Natural climate change processes 71

    Lowland glacial landscapes 77

    Current global climate change: human causes and impacts 83

    Managing global climate change 86

Questions & Answers

Examination skills 92

About this section 93

Q1 Plate tectonics: theory and outcomes 94

Q2 Tropical ecosystems: nature and sustainability 97

Q3 Dynamic coastal environments 100

Q4 Climate change: past and present 104

Knowledge check answers 108

Index 110

■ Getting the most from this book

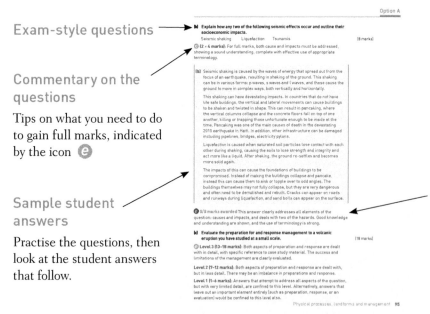

Exam-style questions

Commentary on the questions

Tips on what you need to do to gain full marks, indicated by the icon ⓔ

Sample student answers

Practise the questions, then look at the student answers that follow.

Commentary on sample student answers

Read the comments (preceded by the icon ⓔ) showing how many marks each answer would be awarded in the exam and exactly where marks are gained or lost.

■About this book

Much of the knowledge and understanding needed for A2 geography builds on what you learned for GCSE and AS geography, but with an added focus on the development of specific knowledge of physical geography processes. This guide offers advice for the effective revision of **A2 Unit 1: Physical processes, landforms and management**, which all students need to complete.

The A2 Unit 1 external exam paper tests your knowledge of aspects of physical geography with a particular focus on physical geography processes, landforms and their management. The exam lasts 1 hour 30 minutes and the unit makes up 24% of the final A-level grade.

To be successful in this unit you have to understand:

■ the key ideas of the content
■ the nature of the assessment material — by reviewing and practising sample structured questions
■ how to achieve a high level of performance in the exam

This guide has two sections:

The **Content Guidance** summarises some of the key information that you need to know to be able to answer the examination questions with a high degree of accuracy and depth. In particular, the meaning of key terms is made clear and some attention is paid to providing details of case study material to help to meet the spatial context requirement of the specification. Students will also benefit from noting the exam tips, which will provide further help in determining how to learn key aspects of the course. Knowledge check questions are designed to help learners check their depth of knowledge — why not get someone else to ask you these?

The **Questions & Answers** section includes some sample questions similar in style to those you might expect in the exam. The sample student responses to these questions and detailed analysis will give further guidance in relation to what exam markers are looking for to award top marks.

The best way to use this guide is to read through the relevant topic area first before practising the questions. Only refer to the answers and comments after you have attempted the questions.

Content Guidance

■ Option A Plate tectonics: theory and outcomes

Plate tectonics: margins and landforms

Evidence for the theory of plate tectonics

Structure of the Earth

The Earth has three main sections:

- **Core**: the inner core (which is solid) and the outer core (which is liquid).
- **Mantle**: thickest section of the Earth, around 2,900 km thick.
- **Crust**: oceanic (thinner, but more dense, so heavier) and continental (thicker, but less dense, so lighter).

However, for the purposes of plate tectonics, there are two terms which are most relevant:

- **Lithosphere**: consisting of the crust and upper part of the mantle. This is solid and brittle.
- **Asthenosphere**: consisting of the upper portion of the mantle just below the lithosphere. This is solid but will flow. (This is where the convection currents that drive plate tectonics operate.)

> **Exam tip**
>
> It is good to use appropriate geographical terms in your answers. Refer to lithosphere and asthenosphere rather than crust and mantle.

Knowledge check 1

What are the key differences between lithosphere rock and asthenosphere rock?

According to plate tectonics theory, the Earth's lithosphere is divided into eight major plates and several minor plates. These are moving in relation to one another. Plate tectonics can be seen then as the **theory that outlines processes operating in the lithosphere and asthenosphere which cause the movement of plates across the Earth**.

Evidence for plate tectonics

As global maps became more complete, it was noticed that many of the coastlines of the continents of the world looked almost as though they fitted together like a jigsaw puzzle. In the early twentieth century, Alfred Wegener theorised that all the continents on Earth had been joined together in one massive super-continent known as Pangea. Since then it had broke up and the continents drifted apart. Wegener collated various pieces of evidence in support of his theory.

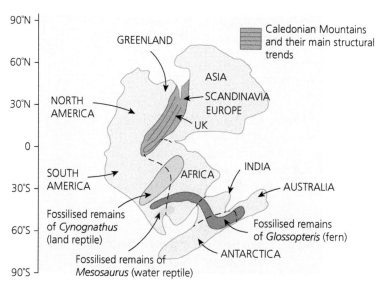

Figure 1 Map of Pangea, summarising the continental evidence for plate tectonics

Geological evidence: rock age, type and structure

Rocks of similar age, type and structure are found in parts of the world separated by thousands of miles. One of the best examples of this is the structural trend lines of the ancient Caledonian Mountains (Figure 1). Not only are the rock types in western Scandinavia, Greenland, Scotland, northwest Ireland, eastern North America and northwest Africa the same, but they have structural trend lines that match up if the continents are positioned side by side on a map.

This implies that the now widely spaced mountain ranges were once side by side. In fact, it is now thought that these mountains were originally part of the Caledonian Mountains found in the centre of the supercontinent Pangea.

Fossil distribution

There are fossilised remains of the same animals and plants located in widely separated continents of the world. For example, remains of the reptile *Mesosaurus* have been found in South America, Africa, India, Antarctica and Australia, now separated by thousands of miles. However, if you join the continents together as Pangea, they form one continuous band. This again implies that the fossils were laid down when the land masses were in very different locations on the surface of the Earth, and that subsequently these land masses have moved thousands of miles apart.

Climatology

In the present day, the fossilised remains of polar plants can be found in India, Australia and southern Africa, and fossils of tropical plants can be found in the mid-latitudes in northern Europe. This implies that these fossils could have been laid down when these continents occupied different positions on the surface of the Earth, and have since moved to their current positions.

All of this evidence suggested that the surface of the Earth was indeed mobile. However, Wegener's theory was not widely accepted, largely because he was unable to

Exam tip

When explaining the pieces of evidence for plate tectonics, show your full understanding by stating both the evidence itself and what that evidence implies.

come up with a process that explained how this movement might occur. It took later evidence from the sea floor to clarify the mechanisms that underpin the theory of plate tectonics.

Ocean floor relief: the mid-ocean ridge and deep ocean trenches

The world's longest mountain range, 75,000 km in length, 1,000 km wide and 2,500 m high, is located as a ridge on ocean floors, and snakes its way around most of the world's major oceans, including the middle of the Atlantic. This ridge has a huge rift valley (canyon) running along its centre. This rift valley was similar to rift valleys in the continents which were pulling apart. This seems to imply that the ocean floor is being split apart at these locations.

Additionally, there are very deep ocean trenches off the coast of some continents, including South America. These trenches show that the ocean floor was sinking down into the asthenosphere at these locations. Together, these facts suggest that new sea floor is being created at the ridges, moving away from there, and being subducted down into the asthenosphere at the trenches. In other words, the sea floor is moving.

Age of the sea floor

As you move out from the mid-Atlantic Ridge towards the continents surrounding the Atlantic Ocean, the rocks get progressively older, again following a symmetrical pattern either side of the ridge. This consolidates the idea of the sea floor moving: splitting apart at the mid-ocean ridges, moving away from the ridges, and sinking down into the asthenosphere at the trenches (Figure 2).

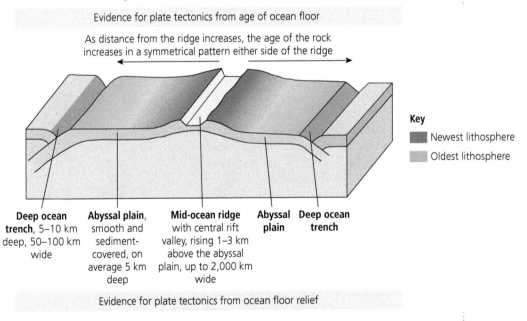

Evidence for plate tectonics from age of ocean floor

As distance from the ridge increases, the age of the rock increases in a symmetrical pattern either side of the ridge

Key

◼ Newest lithosphere

◼ Oldest lithosphere

Deep ocean trench, 5–10 km deep, 50–100 km wide

Abyssal plain, smooth and sediment-covered, on average 5 km deep

Mid-ocean ridge with central rift valley, rising 1–3 km above the abyssal plain, up to 2,000 km wide

Abyssal plain

Deep ocean trench

Evidence for plate tectonics from ocean floor relief

Figure 2 Sea floor evidence for plate tectonics

Palaeomagnetism

In the late 1950s, scientists had recently discovered that the Earth's magnetic pole periodically flips or reverses. Instead of being at the north as it is now, it moves to the

South Pole, and then back to the north, over and over again. As magma cools, the iron within it orientates itself towards the direction of the magnetic pole, pointing to the position the magnetic pole was in when the magma was cooling.

During the 1960s, as ships with magnetometers were mapping the ocean floors, a particular pattern began to be revealed. As you move out from the mid-Atlantic Ridge, there is a symmetrical striped pattern of magnetic orientation found within the rocks. What the magnetometers revealed was that there were alternating bands of different magnetic orientation, first pointing north, then south, then north again and so on, and that these bands were symmetrical either side of the ridge.

This suggested that iron in the magma erupting from undersea volcanoes orientated itself to the north as it cooled. Then the magnetic pole flipped to the south and the iron in the new magma that erupted orientated itself south and displaced the previous cooled magma away from the ridge.

The global pattern of earthquakes

During the Cold War, a network of seismometers was established to monitor underground nuclear testing. A positive by-product of this was that, for the first time, scientists were able to observe the global pattern of the locations of the world's largest earthquakes. This revealed a clustering of major earthquakes following a linear pattern: a series of narrow bands across the lithosphere. Studies on the direction of movement of the lithosphere at these narrow bands revealed that the lithosphere was broken up into different sections (known as plates) which were moving independently of one another: sometimes away from each other, sometimes towards each other and sometimes alongside.

Exam tip

For both ocean floor relief and age of the sea floor, it is important to point out that the patterns are symmetrical either side of the ridge.

Exam tip

If an exam question gives you a choice of which evidence to discuss, if you choose evidence from the sea floor, this will allow you to discuss more directly the evidence for the mechanism of plate tectonics.

The theory of plate tectonics

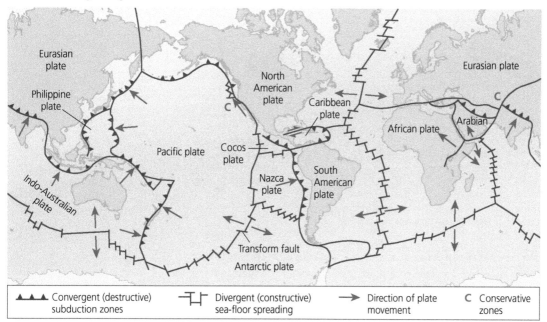

Figure 3 Global map of plates and their margins

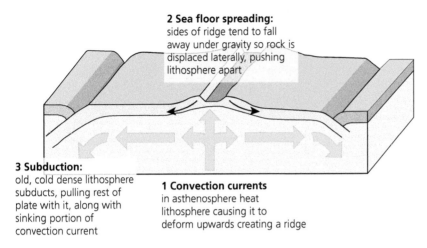

2 Sea floor spreading: sides of ridge tend to fall away under gravity so rock is displaced laterally, pushing lithosphere apart

3 Subduction: old, cold dense lithosphere subducts, pulling rest of plate with it, along with sinking portion of convection current

1 Convection currents in asthenosphere heat lithosphere causing it to deform upwards creating a ridge

Figure 4 The mechanisms of plate movement

Processes at plate margins

Constructive margins

At constructive margins (Figure 5), plates move apart. An example is the mid-Atlantic Ridge.

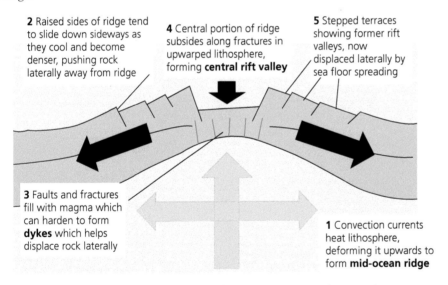

2 Raised sides of ridge tend to slide down sideways as they cool and become denser, pushing rock laterally away from ridge

4 Central portion of ridge subsides along fractures in upwarped lithosphere, forming **central rift valley**

5 Stepped terraces showing former rift valleys, now displaced laterally by sea floor spreading

3 Faults and fractures fill with magma which can harden to form **dykes** which helps displace rock laterally

1 Convection currents heat lithosphere, deforming it upwards to form **mid-ocean ridge**

Figure 5 Processes and landforms at a constructive margin

Sub-plate processes

The sub-plate processes at constructive margins are dominated by **convection currents**. Heat escaping from the core and moving out through the mantle heats the rock in the asthenosphere, causing it to flow upwards towards the lithosphere. When this rising part of the convection current reaches the bottom of the lithosphere, it heats it and becomes less dense, and stretches and deforms upwards.

Plate processes and resultant landforms

The sub-plate processes are dominated by **sea-floor spreading**, a process that explains how plates are in effect pushed apart at constructive margins. How does this movement occur? First, as the lithosphere is heated by the convection current below, it becomes less dense and is stretched and deformed upwards to form an **ocean ridge**. This is a raised portion of the sea bed; it may be many thousands of kilometres in length and 1 to 3 km in height. It can also be very wide: the mid-Atlantic Ridge is up to 2,000 km in width. As the ridge begins to rise above the **abyssal plain**, it becomes increasingly rugged as you move towards its centre. It also has a series of eroded sub-ridges running parallel to the main ridge.

The centre of the ocean ridge often features a **central rift valley**. This linear valley runs along the length of the ridge; it can be as much as 1.5 km deep and up to 30 km wide. The inner slopes of the rift valley are steep and relatively straight. The outer slopes have a much gentler gradient and are marked by a series of stepped terraces (the edges of former rift valleys) parallel to the main ridge.

Over time, the raised sides of the ridge tend to slide down sideways away from the centre of the plate margin under the influence of gravity as they cool slightly and become more dense. Secondly, the fracturing of the rock during this deformation allows magma to intrude into the lithosphere, where it may cool and harden to form **dykes** (vertical intrusions of igneous rock). This intruding magma can help displace the rock laterally away from the margin.

Conservative margins

At conservative margins (Figure 6), plates move alongside each other. An example is the San Andreas Fault.

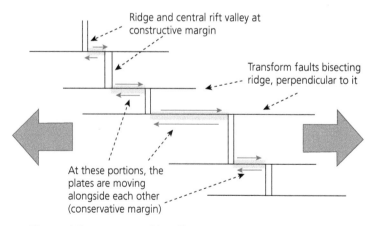

Ridge and central rift valley at constructive margin

Transform faults bisecting ridge, perpendicular to it

At these portions, the plates are moving alongside each other (conservative margin)

Figure 6 Processes and landforms at a conservative margin

Plate processes

Conservative margins are places where plates move alongside each other. They result from the often very complex way in which plates move as they curve across the ocean floor. So, to avoid the margin twisting into an 'S' shape, instead it fractures perpendicular to the margin. This creates a series of **transform faults** (also known as **conservative margins**) bisecting the constructive margin, forming a zig-zag pattern along the length of the ridge.

Most conservative margins are under the oceans. But the most famous one in the world – the San Andreas Fault – runs through much of the length of the state of California. It is a transform fault that connects the ridge marking the constructive margins between the North American, Pacific and Juan de Fuca plates.

Sub-plate processes

The sub-plate processes occurring here are similar to the ones we have covered above. For instance, if you are referring to a conservative margin in the form of a transform fault bisecting a constructive margin, then the sub-plate processes are the **convection currents** we have referred to above.

Collision margins

At collision margins, plates move together (Figure 7). An example is the Himalayas.

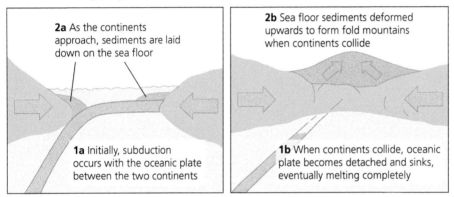

2a As the continents approach, sediments are laid down on the sea floor

1a Initially, subduction occurs with the oceanic plate between the two continents

2b Sea floor sediments deformed upwards to form fold mountains when continents collide

1b When continents collide, oceanic plate becomes detached and sinks, eventually melting completely

Figure 7 Processes and landforms at a collision margin

Sub-plate processes

Initially, as the two continents approach one another, the oceanic lithosphere between them subducts below one of the continents (at this stage it is very like a destructive margin). Eventually, the two sections of continental lithosphere collide and the subducting oceanic lithosphere becomes detached from the continents above. It continues down into the asthenosphere where it slowly melts and becomes assimilated fully into the asthenosphere. Thus, eventually all subduction stops below a collision margin.

Plate processes

As the two plates converge, sediments are laid down on the sea floor. These come from two main sources: first, sediment is deposited on the sea bed by rivers flowing off the edge of the continents. Secondly, as the oceanic lithosphere subducts, much of the sediment that has been deposited on top of it is scraped off and builds up on the surface (this sedimentary material is less dense than the basaltic oceanic plate and so resists subduction).

When the two continents meet, the continental lithosphere is too buoyant to subduct and so the two plates collide into each other and the lithosphere is compressed. As a result, the sea floor sediments are folded upwards creating **fold mountains**. The process of deformation causes multiple faults and fractures in the mountains. These mountains also have a deep 'root' extending well below the surface.

Destructive margins

At destructive margins, plates move together (Figure 8). An example is the western coast of South America.

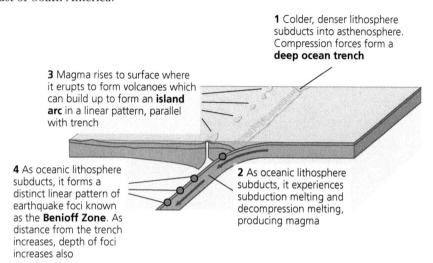

1 Colder, denser lithosphere subducts into asthenosphere. Compression forces form a **deep ocean trench**

3 Magma rises to surface where it erupts to form volcanoes which can build up to form an **island arc** in a linear pattern, parallel with trench

4 As oceanic lithosphere subducts, it forms a distinct linear pattern of earthquake foci known as the **Benioff Zone**. As distance from the trench increases, depth of foci increases also

2 As oceanic lithosphere subducts, it experiences subduction melting and decompression melting, producing magma

Figure 8 Processes and landforms at a destructive margin

Plate processes

At destructive margins, the oceanic lithosphere is often many thousands of miles away from the place where it was created and by now it is cold and therefore denser. This means that the lithosphere will tend to **subduct** back down into the asthenosphere. Where oceanic lithosphere meets continental lithosphere, the more dense oceanic plate will subduct below the continental plate. In fact, continental lithosphere is not dense enough to subduct into the asthenosphere. When two plates consisting of oceanic lithosphere meet, the lithosphere that is colder and denser subducts.

Sub-plate processes

The sub-plate processes at destructive margins are dominated by **subduction**, aided by the sinking of the **convection currents** in the asthenosphere. The oceanic lithosphere subducts down into the asthenosphere at an angle typically of around 45 degrees; this subducting segment is known as the **Benioff Zone**. As it subducts, the surface of the lithosphere melts as it comes into contact with the hotter asthenosphere (**subduction melting**). In addition, sea water carried down by the subducting plate not only lowers the melting point of the lithosphere, but then mixes with the melted material, reducing its viscosity and helping it flow more easily (**hydration melting**). This melting begins at a depth of around 80 km. By depths of around 600–700 km, subduction stops as the descending oceanic lithosphere has been fully assimilated into the asthenosphere.

The process of subduction is aided by sinking **convection currents** in the asthenosphere. These convection currents themselves now are cooler and denser. As they sink, they help drag the cool, dense oceanic lithosphere down with them.

Exam tip

If discussing destructive margins in the exam, make sure that you explicitly use the terms **subduction** and **convection currents** in your answer.

Resultant landforms at destructive margins

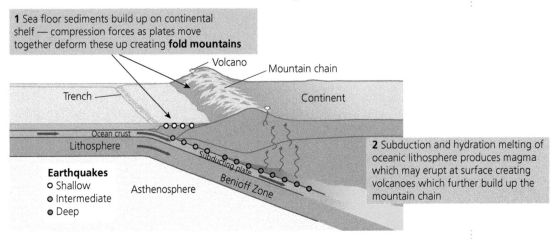

1 Sea floor sediments build up on continental shelf — compression forces as plates move together deform these up creating **fold mountains**

Volcano — Mountain chain

Trench —

Continent

Ocean crust

Lithosphere

Subducting plate

Earthquakes
o Shallow
⊙ Intermediate
● Deep

Asthenosphere

Benioff Zone

2 Subduction and hydration melting of oceanic lithosphere produces magma which may erupt at surface creating volcanoes which further build up the mountain chain

Figure 9 Landforms at a destructive margin

The two main landforms formed here are fold mountains, and deep sea trenches and island arcs (Figure 9).

Fold mountains: these are formed in a similar way to those at collision margins (see p. 12). In addition, these mountain ranges may be further built up by **volcanoes** (formed by the melting of the subducting plate to produce magma that rises through the lithosphere above to erupt at the surface). All 25 of the world's volcanoes over 6,000 m in height are found in the Andes.

Deep sea trenches and island arcs: the **deep sea trenches** that are found here form a linear pattern, often many thousands of kilometres long. They are typically between 50 and 100 km wide and 5–10 km deep (the deepest trench in the world, the Marianas Trench, reaches just over 11 km in depth).

Also forming a linear pattern, running parallel to the trench on the non-subducting plate, is the **island arc** (see Figure 8). These islands are usually between 100 and 200 km from the trench (the distance depends on the angle of Benioff Zone in the subducting plate). They consist of volcanic cones and the rock is made up of cooled andesitic lava (see later notes on volcanic activity).

When two oceanic plates meet, the one that is colder and thus more dense will subduct. As subduction occurs, compression forces cause the plate to buckle and deform, producing the deep sea trench as the leading edge of the non-subducting plate is deformed downwards.

As the plate subducts down further into the hotter asthenosphere, its surface begins to melt. This melting process is aided by the sea water that was brought down with the plate during subduction. The water lowers the melting point of the lithosphere. This melting begins at a depth of around 80 km (without the water to aid the melting process, it would not begin until a depth of around 200 km). This partial melting creates magma which is less dense than the surrounding asthenosphere. It rises upwards through the lithosphere and erupts at the surface to form volcanoes. Over time, these volcanoes may build upwards to emerge above the ocean surface, forming the arc of islands, parallel with the trench.

Summary

- There are various pieces of evidence that point towards the theory of plate tectonics, including: geologic evidence (age, type and structure of rock), fossil evidence and evidence from climatology; ocean floor relief, age of the sea floor, and the global pattern of earthquakes.
- The theory of plate tectonics states that the Earth's lithosphere is broken up into a number of plates that move. This movement is caused by the interaction of sea floor spreading and subduction, driven by convection currents in the asthenosphere.
- These processes operate at various margins (constructive, destructive, conservative and collision) to produce various landforms (ocean ridges and rift valleys, deep sea trenches and island arcs, and fold mountains).

Volcanic activity and its management

Volcanic activity at plate margins and hot spots

Constructive and destructive margins

Table 1 Volcanic activity at constructive and destructive margins

	Constructive margins	Destructive margins
Cause	Magma produced by partial melting of asthenosphere resulting from fall in pressure as convection current rises (**decompression melting**).	As the oceanic lithosphere subducts, it experiences partial melting as it enters the hotter asthenosphere (**subduction melting**), a process aided by the sea water carried down during subduction (**hydration melting**).
	In both cases, this magma is less dense than the asthenosphere so rises to fill underground **magma chambers**, from where it may rise to the surface and erupt as **lava**.	
Magma/lava	**Basaltic magma** Quite runny and so gases within it can more easily escape.	**Andesitic** or **rhyolitic magma** is produced. Very viscous so gases within it cannot escape as easily and tend to be trapped within it.
Nature of eruptions	Rarely explosive as gases less likely to be trapped in magma.	Explosive eruptions common. Trapped gas expands, blowing magma apart.
	Eruptions dominated mostly by **lava flows**.	Some **lava**. Explosive materials, e.g. tephra, ash, pyroclastic flows (see pp. 17–18).
	Eruptions more frequent and continuous.	Eruptions less frequent, as vents can become blocked by the more viscous magma.
Form of volcano	**Shield volcanoes** More gently sloping but cover large areas, as the less viscous lava can flow easily and cover larger areas. Lava plateau / Shield volcano / Lava only	**Composite volcanoes** Steeper and more localised, as viscous lava does not flow as fast and tends to cool closer to volcano. Acid lava dome / Composite cone / Column of gas and finer fragments / Layers of lava/ash

Knowledge check 3

Outline the various processes by which magma can be produced.

Hot spots

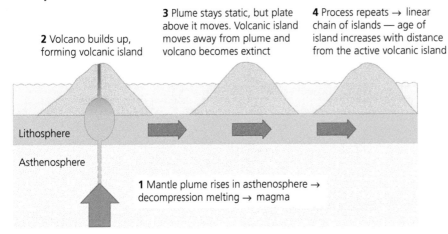

Figure 10 Volcanic activity at hot spots

The Hawaiian Islands are a good example to use to understand the nature of and processes associated with **hot spot** activity (Figure 10). The Hawaiian Island chain forms a **linear pattern** of islands, running 500 km northwest away from the main island of Hawaii. All the islands are made of basaltic lava and have volcanoes on them, but the only active volcanoes are found at the southeastern end of the chain on the main island. In addition, the age of the islands increases as you move from the main island in the southeast up through the islands towards the northwest.

Processes

This pattern can be explained as follows. The volcanic activity is produced by a **mantle plume**, a convectional flow of heated rock in the asthenosphere. As this mantle plume rises, magma is produced by partial melting of the asthenosphere resulting from a fall in pressure (decompression melting). This magma is less dense than the asthenosphere and is hot enough to melt the lithosphere, and so it rises to erupt at the surface, forming volcanoes. Further eruptions cause the volcano to rise above the sea surface, forming a volcanic island directly over the hot spot.

The mantle plume stays static in its position. However, the Pacific plate above it is moving from the southeast to the northwest and so this island gradually moves away from the hot spot and the volcano becomes extinct. As the plate continues to move, a new portion of lithosphere is now over the hot spot, and the same processes form a second volcanic island. Over time, it too moves away from the hot spot, and the whole process repeats over and over again, forming this linear pattern of islands with only the last island in the chain to the southeast having active volcanoes on it.

Exam tip

It is important to note that the pattern formed by the Hawaiian Islands is a **linear** one, with only the island at the end containing active volcanoes.

Knowledge check 4

What is the pattern of ages of rocks associated with the islands produced by the Hawaiian hot spot?

Exam tip

Make sure you state that, at a hot spot, the mantle plume stays static. It is the plate that moves above it.

Socioeconomic and environmental hazards and benefits of volcanic activity

Hazards

Lava flows

Lava flows can be fast moving — viscous lava coming out of steeper volcanoes can move at speeds up to $14\,\mathrm{m\,s^{-1}}$ — or slow moving viscous lava.

Socioeconomic hazards: slow moving lava tends not to threaten human life. However, it can still destroy property and farm land, and have economic impacts, for example the 1973 eruption in Heimaey, Iceland.

Environmental hazards: in the short term, these lava flows can completely devastate the natural environment. All vegetation is covered completely, destroying it and the habitats it provides for animals.

Pyroclastic flows

Pyroclastic flows are explosive eruptions of superheated tephra and poisonous gases. They are directed laterally and surge downhill, travelling at speeds of over $30\,\mathrm{m\,s^{-1}}$ and may travel distances of up to 30–40 km from the volcano at temperatures that may reach 1,000°C. As they travel down river valleys on the sides of volcanoes, they can even surge up the valley sides, engulfing all and everything in their path.

Socioeconomic hazards: pyroclastic flows have been responsible for more volcanic related deaths than any other hazard, e.g. Pompeii, 79 AD. They can have a devastating economic impact also. They may be preceded by an air blast that can topple buildings. The heat can set fire to buildings.

Environmental hazards: in the short term, pyroclastic flows can completely devastate the natural environment. All vegetation is covered completely, destroying it and the habitats it provides for animals, e.g. Mt St Helens, 1980.

Lahars

Lahars are volcanic mudflows that occur when volcanic material, such as ash and rock, mixes with water, including heavy rain, melted snow from a volcanic peak, and collapse of a crater lake in a volcano. Lahars form viscous flows and move downhill, following river valleys. The largest ones can reach speeds of up to $100\,\mathrm{km\,h^{-1}}$. When they stop, they quickly set and solidify, burying anything in their path.

Socioeconomic hazards: lahars are the second biggest cause of volcanic deaths after pyroclastic flows, e.g. Nevado del Ruiz, 1985, and can cause significant economic losses as they damage and destroy buildings and infrastructure, and crops and livestock.

Environmental hazards: lahars can significantly impact the environment in the short term as they will bury the land in deposits of mud, metres in depth, e.g. Nevado del Ruiz, 1985.

Tephra and ash clouds

If a volcano erupts explosively, it will generate tephra: the largest tephra are in the form of volcanic bombs which are forcefully thrown out into the area close to the volcano. The smallest material, **volcanic ash**, is thrown high into the atmosphere

> **Exam tip**
>
> Ensure you can make reference to a place for each of the hazards of volcanic activity. You will need to be able to refer to places to get a top-level answer.

> **Exam tip**
>
> In the late 1980s, scientists produced a video of the various volcanic hazards which they used to educate people and authorities in volcanic areas over the world to increase their awareness of the dramatic nature of these hazards. You can do the same using YouTube. Search for videos of each of the examples given above to deepen your understanding.

where it may be transported by air currents many hundreds or even thousands of miles, being deposited over very large areas.

Socioeconomic hazards: deaths from volcanic ash are rare, but local conditions can make the ash more hazardous, e.g. in Mt Pinatubo. The ash there caused some buildings to collapse, resulting in economic loss and deaths. The 2010 eruption of Eyjafjallajökull in Iceland sent fine ash all over Europe, grounding planes and causing significant economic losses.

Environmental hazards: the 1991 Mt Pinatubo eruption threw so much ash high up into the atmosphere that temperatures across the world were affected. The summer temperatures in Europe the following year were around 2°C lower than average.

Poisonous gas

Magma contains many toxic gases such as carbon dioxide (CO_2) and hydrogen sulphide (H_2S). During eruptions, these can be released high into the atmosphere. The gases can be released at times other than an eruption, which can present a considerable hazard to people living in the area.

Socioeconomic hazards: carbon dioxide escaped from the volcanic Lake Nyos in Cameroon in 1986. It sank down the slopes of the volcano, killing 1,700 people and over 8,000 livestock.

Environmental hazards: the slow release of carbon dioxide and sulphur dioxide (SO_2) around a volcano can destroy plants, e.g. Mammoth Mountain, USA.

Jökulhlaup

This is a flood that is caused by glacial melting associated with a volcanic eruption.

Socioeconomic hazards: as with any flood, the waters can cause significant damage to infrastructure e.g. Grimsvötn, Iceland, 1996. However, due to close monitoring of the hazard, people were evacuated in time.

Environmental hazards: the 1996 jökulhlaup at Grimsvötn caused significant subsidence of the glacial lake.

Socioeconomic benefits

Fertile soils supporting agriculture

In the longer term, the ash and tephra emitted by volcanoes improves soil fertility in the areas affected by them. For example, most of Italy has poor soil with lots of limestone. In contrast, in the regions around Naples, the soils are more fertile as a result of various eruptions in the past from Mt Vesuvius. This increases agricultural productivity.

Mineral creation and extraction

Volcanic materials are mined for commercial purposes. For example, pumice is used to act as abrasives in soaps and household cleaners. Ash and pumice can be used for aggregate in the making of cement. Valuable metals, such as gold and uranium, along with precious stones including opals and onyx, are also found in volcanic areas. At the Ijen volcano in Indonesia, workers mine the sulphur deposits found there. These can all boost the economy of the country with the volcano and provide employment for locals.

Knowledge check 5

Which volcanic hazards pose the greatest threat to human life and why?

Exam tip

Do an internet image search for each of the benefits. Exam questions can give you photographs of them and ask you to identify them.

Tourism

Volcanoes naturally draw tourists, and volcanic tourism can help boost the economies of the countries that can offer this attraction. For example, Mt Vesuvius in Italy and the nearby well-preserved ruins of Pompeii attract 400,000 and 2 million annual visitors respectively. It is not just active volcanoes that attract tourists either, The 60-million-year-old Giant's Causeway in Ireland attracted over 850,000 visitors in 2015, bringing a boost to the economy, providing jobs for locals and helping support the conservation work of the National Trust there.

Environmental benefits

Land creation

Volcanic eruptions can create new land, for example the volcanic islands of Hawaii and the Antilles archipelago in the Caribbean. Where these islands formed, unique plants and animals adapted to living there, increasing global biodiversity.

Geothermal energy

The heat rising in the crust in volcanic areas can be harnessed for power generation. Countries like Iceland, Costa Rica, New Zealand and Kenya can generate significant amounts of their energy needs from geothermal sources — ranging from 14% in Costa Rica to 51% in Kenya. The processes involved in generating geothermal energy release a small fraction of the carbon dioxide released from conventional fossil fuel plants. So it is reliable, sustainable and relatively non-polluting: in a world experiencing climate change this can make a positive contribution to the challenges of providing sustainable energy.

> **Exam tip**
>
> Ensure you can make reference to a place for each of the benefits of volcanic activity.

Evaluation of how a country prepares for and responds to volcanic activity

Unlike earthquakes (see pp. 24–25), volcanoes may give off a number of warning signs that an eruption might be about to happen. Although very few of the world's active volcanoes are being monitored on an ongoing basis, if these warning signs start to appear, many countries will be able to send in scientists to begin more detailed monitoring.

> **Exam tip**
>
> The warning signs include things such as increasing earthquake activity (resulting from the magma intruding into the ground), emissions of steam, increases in the amount of volcanic gases (such as sulphur dioxide and carbon dioxide) being emitted, deformation of the land (it may bulge outwards as magma builds up underground), alterations in gravity and magnetism due to the rising magma.

Mt Pinatubo, 1991

After being dormant for over 500 years, the Mt Pinatubo volcano erupted into life again in a huge eruption in 1991.

Preparation: monitoring

- 2 April 1991, steam emissions from side of volcano; scientists from PHIVOLCS arrived and began on-site monitoring
- 10 km radius danger zone declared around the volcano
- USGS team arrived with seismometers, tilt meters and other equipment:
  - They monitored the current activity on the volcano (around 40–150 earthquakes were occurring each day, and sulphur dioxide levels rose from 500 tonnes per day to 5,000 tonnes per day in five weeks.
  - They conducted field studies to examine the eruption history of the volcano, looking for evidence of where previous lahars and pyroclastic flows had travelled (they discovered that the volcano had only erupted four or five times in the past 2,000 years, indicating that an eruption here was likely to be explosive in nature).

PHIVOLCS: Philippines Institute of Volcanology and Seismology

USGS: United States Geologic Survey

Preparation: hazards map

The scientists produced a hazards map (Figure 11), indicating the likely nature and extent of the eruption. And, as earthquake frequency increased and sulphur dioxide levels decreased slightly (showing the magma was holding on to its gas, meaning that the eruption would be highly explosive), phased evacuations were ordered. On 3 June, 20,000 people living within 10 km were evacuated; by 7 June, a further 120,000 people within 18 km were evacuated. The volcano erupted on 15 June.

Evaluation

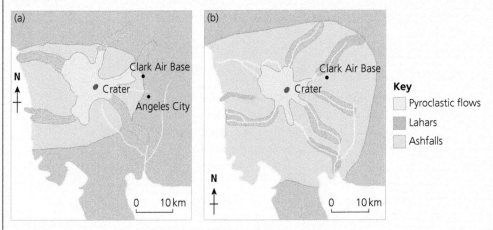

Figure 11 (a) Hazard prediction map and (b) a map of the actual impacts of the Pinatubo eruption

There are both positive and negative aspects of the preparations for the eruption:

Positive

- The hazard maps proved largely accurate — all the areas that were predicted to be affected were so.
- Relatively small numbers of people were killed (around 850) for the massive size of the eruption and the numbers of people living nearby (around 1 million).
- USGS/PHIVOLCS estimate:
  - 5,000 to 20,000 lives were saved due to the predictions and evacuations.
  - Only 20 of the 20,000 Aeta highlander people on the volcano were killed.
- The cost of the monitoring was very low at around $1.5 million and the costs of the evacuations were around $40 million.

- This is less than expected as the economic losses were averted, especially as aircraft were moved out of the way.
  - USAF — $200 million in losses were averted.
  - Philippine and commercial airlines saved between $50 and $100 million also.

Negative

- The hazard map underestimated the extent of the hazards. For example, the map predicted ash would fall over a radius of around 40 km to the east of the volcano — in fact, it travelled out about 65 km in places.
- There was the unfortunate coincidence of Typhoon Yunya. The heavy rains increased the weight of the ash on the buildings, causing more damage and collapse than had been forecast.

Overall, thanks to the efforts of the scientists and authorities, this massive eruption had a relatively small impact on human lives, and so the preparations were very effective.

Response: immediate

There are both positive and negative aspects of the immediate response to the eruption:

Positive

- The National Disaster Coordinating Council mobilised to aid in the evacuation, rescue and relief response. This included providing social services (health and basic education) in the evacuation centres.
- Health advisories were also issued to inform people how to deal with the ash. The fine particles in the ash could be an eye irritant and could trigger asthma attacks.

Negative

- The situation for the people who were evacuated was far from perfect. Disease was a major issue in the evacuation centres. Some estimates suggest that around 600 of the 850 deaths resulting from the eruption were due to disease caused by poor conditions in the evacuation shelters. For example, 30,000 people used the Amoranto Velodrome in Quezon City for temporary shelter. Many others relocated to the shanty towns in cities like Angeles where, again, disease spread, affecting hundreds of people.

Overall, the response did help support a large number of the evacuees. However, disease in the inadequate evacuation shelters claimed more lives in the aftermath of the eruption.

Response: longer-term

There are both positive and negative aspects of the longer-term response to the eruption:

Positive

- A system to monitor (rain gauges, acoustic monitoring to detect the vibrations of passing lahars, and manned watchpoints) and warn of lahars was rapidly established. This system has saved hundreds of lives by enabling warnings to be sounded for most but not all major lahars at Pinatubo since 1991.
- In the longer term, many of the evacuees needed to be resettled and helped to re-establish their livelihoods. By June 1991, the government had set up the Task Force Mt Pinatubo, a unit with a six-year mandate and a 10 billion peso fund to assist in: the establishment of resettlement centres; providing employment and livelihood opportunities for victims, repair of damage infrastructure; and providing relief funds for victims. By 1996, all government departments included post-eruption mitigation strategies in their programmes.

Negative

- The efforts by engineers to tap lahars behind dams and levees have largely failed.
- Dozens of people have been killed since 1991 by lahars, especially on the Pasig-Potrero river where three major lahars occurred (1991, 1992, 1994).
- The homes of 100,000 people have been destroyed by lahars since the eruption.

Overall, despite the challenges of the lahars, the actions taken have been largely effective, as many hundreds of lives have been saved.

Summary

- Volcanic activity can be found at constructive margins, destructive margins and hot spots.
- There are a number of socioeconomic and environmental hazards associated with volcanic activity, including: lava flows, pyroclastic flows, lahars, ash and poisonous gas.
- However, volcanoes can bring some socioeconomic and environmental benefits also,

including fertile soil, mineral creation, tourism, land creation, and geothermal energy.
- Volcanoes, unlike earthquakes, can give warning signs of imminent eruptions. However, these warning signs may not indicate an eruption will occur, and so prediction of volcanoes still has its challenges.

Seismic activity and its management

Nature and impact of seismic events

Earthquakes

An earthquake is the shaking of the earth resulting from the sudden release of built-up seismic strain energy along a fault. When stress is placed on rock due to the movement of the lithosphere, at first that rock will slowly deform. This process of slow deformation may take years, decades or even centuries and all the while strain energy is being stored. Eventually, the strain on the rock becomes too much, and it suddenly snaps back into its original shape, but in a new position relative to where it started. This sudden release of the stored-up energy causes the ground to shake and **waves** are sent out from this point of origin, known as the **focus**. The point on the Earth's surface directly above it is known as the **epicentre**.

P, S and L waves

It is possible to identify various forms these waves can take (Table 2).

Table 2 P, S and L waves

Body waves Will travel through the body of the Earth in all directions, including into its interior		Surface waves Travel only through the lithosphere
P waves (primary waves)	**S waves** (secondary waves)	**L waves** (love waves)
Compressional wave, pushing and pulling the substance it is travelling through	Distortion wave, causing movement up and down or side-to-side, perpendicular to the direction of travel	Distortion wave, causing side-to-side movement perpendicular to the direction of travel

Table 2 *continued*

P waves (primary waves)	S waves (secondary waves)	L waves (love waves)
Travel at around 6 km s⁻¹ (more than seven times the speed of sound) in the lithosphere, and up to 13 km s⁻¹ through the core	Travel at around 3.5 km s⁻¹	Travel at around 90% of the speed of S waves
Travel through solid and liquid (so will travel through the Earth's core)	Do not travel through liquid (so will not pass through water, molten rock, or the Earth's outer core)	Travel only through the lithosphere
Cause little damage to infrastructure	Cause more damage to infrastructure	Cause most damage to infrastructure

Earthquakes at different plate margins

Earthquakes mostly tend to have shallow foci at the following margins: constructive, conservative and collision. However, at destructive margins, there are both shallow and deep focus earthquakes and they have the following pattern: as distance from the trench increases along the Benioff Zone, the depth of earthquake foci increases. The reason for this is that the earthquake is associated with the release of strain energy along the subducting plate. Thus, the foci get deeper as the plate subducts further down (see Figure 8 on p. 13).

The impacts of earthquakes

Seismic shaking

Seismic shaking resulting from an earthquake can cause devastation to infrastructure. In the 2010 earthquake in Haiti, for example, the buildings were very poorly constructed with under-reinforced vertical columns. This resulted in **pancaking** — where the vertical concrete columns of the buildings give way during the twisting of the earthquake, causing the concrete floors to collapse flat, one on top of the other (ending up layered like a pancake stack). Clearly such catastrophic failure can have devastating consequences for anyone in the building at the time: many will be killed outright; those who survived are often badly injured and trapped in the rubble (with, of course, the constant threat of aftershocks hanging over them).

In light of this, there has been much emphasis in more developed countries to make buildings in earthquake-prone regions **life safe**. For example in Japan, strict building regulations exist (see Case study on pp. 26–28).

The intensity of shaking is only partly determined by the magnitude of the earthquake. One other major factor at play is the nature of the land on which the buildings are built. If the land is made up of sediments rather than rock, the sediments can actually amplify the ground shaking during an earthquake and thus the damage will be greater than that experienced by ground underlain by solid rock during the same earthquake.

Liquefaction

Liquefaction occurs when a saturated or partially saturated soil loses its strength and integrity as a result of the shaking that occurs during an earthquake. During the shaking, water pressure increases and the sand particles in the soil lose contact with

Exam tip

Make sure you are clear on the distinction between the three types of waves and can explain each one clearly in the exam.

Knowledge check 6

'Earthquakes don't kill people, buildings do.' To what extent is this cliché true?

each other. As a result, the soil begins to flow and move more like a liquid. When the shaking stops, the water pressure falls again, and the soil becomes solid once more.

This occurs especially in more saturated, lower density sandy soils. These can often be found on reclaimed land close to coastal areas.

Liquefaction can cause various serious impacts, including:

■ The foundations of buildings and infrastructure can become compromised, causing the lower levels to sink down, or the whole building to tilt at an unusual angle. These buildings may be beyond repair following the earthquake.
■ Underground pipes and cables may be damaged.
■ The upward water pressure may allow water to enter buildings via service ducts and cause water damage to the building.
■ Sloping ground may slide along a liquefied soil layer, causing large cracks and fissures to appear in the ground. These can damage any human structures sitting upon them.
■ Manhole covers may be lifted up above the road level by the liquefied soil.

Tsunamis

Tsunamis occur when earthquakes on the ocean floor cause vertical displacement of the lithosphere. For example, at a destructive margin, the leading edge of the non-subducting plate is dragged down during subduction. However, when it springs back upwards, it displaces the column of water above it vertically, creating tsunami waves on the ocean surface. These waves fan out from the epicentre.

In deep water, tsunamis are very fast, with very low wave heights that may be less than a metre, but with very long wavelengths that might be measured in kilometres. As the waves approach the shore, however, the leading wave crest slows down due to friction with the shallower sea bed and the wave crest behind it starts to build up. This reduces the wavelength, but can lead to dramatic increases in wave height to tens of metres.

Tsunamis can have devastating impacts. The waves can travel many kilometres inland, bringing devastating destruction to infrastructure (both through the power of the wave, and via the debris of buildings already destroyed). Without adequate warnings, hundreds of thousands of lives can be lost, as with the Boxing Day tsunami in southeast Asia in 2004. But even with warnings, large tsunamis can kill tens of thousands (see Case study on pp. 26–28). Tsunamis can also devastate coastal ecosystems, coral reefs and mangrove swamps.

Attempts to predict seismic events

Seismic gap theory

A **seismic gap** is a section of fault that has not moved seismically over an unusually long time period. This theory is based on the hypothesis that, given enough time, all parts of an active fault will move. If most parts of a fault have already moved, the section that is still yet to move is the place where the next large earthquake is most likely.

The most quoted example of the seismic gap theory in practice was the Loma Prieta earthquake on the San Andreas Fault in California in 1989. Although there had been seismic movement along most sections of the fault, by the mid-1980s, three sections had not moved in over 20 years: San Francisco, Parkfield and Loma Prieta. When the fault moved in 1989, it filled the Loma Prieta gap.

However, the theory has its critics. Just because one section has not moved, it does not follow that it will be the next section to move, as there may be other factors at work restricting movement. For example, the unbroken stretch of fault may be stronger than those surrounding, or the movement in the nearby section of faults may have relieved the strain on the seismic gap.

One example of a location where a seismic gap has not yet been filled is the Tokai region in southwest Japan. This section of fault is overdue movement if you look at both the movement of the adjacent sections of fault and historic records of earthquake intervals. However, the magnitude 9 earthquake in 2011 did not occur here, but on the Tohoku fault to the northeast of the country.

Dilation theory

Another area of research has focused on looking for precursors to major earthquakes, advance warning signs that indicate an earthquake is about to happen. One of these is the **dilation theory**. Observations of rocks undergoing stress reveal that the rock begins to dilate (expand) as a result of microcracks that develop in the rock. This dilation has a number of consequences that may be observed, including change in the speed of seismic waves travelling through the rock, changes in its magnetic and electrical resistance properties, and increases in the release of gases such as radon. The theory proposes that, if recognisable patterns in these things can be observed prior to an earthquake happening, then it may be possible to use them to predict future earthquakes.

This theory was put to the test in Parkfield, California, following the Loma Prieta earthquake in 1989. Parkfield lies on the San Andreas Fault and it had been experiencing earthquakes at reasonably regular intervals of around 20 years, with the previous one being in 1966. Seismologists arrived with many instruments to monitor the area, looking for evidence of dilation. In fact, not only did they have to wait a full 15 years until 2004 for a medium-sized earthquake, but that quake produced no precursors at all.

At present, therefore, there are no ways of accurately predicting when an earthquake may happen. The best that can be done is to state the probabilities of earthquakes in the future. For example, the USGS has calculated that there is a 62% chance of an earthquake magnitude 6.7 or greater in the San Francisco Bay area between 2003 and 2032.

Evaluation of how a country prepares for and responds to seismic activity

If we are currently unable to forecast earthquakes accurately, then countries at risk from major quakes need to prepare well in advance through things like building codes to make buildings life safe, educating people to know what to do in the event of a major quake, developing earthquake early warning systems and building tsunami walls. They must also be ready to respond quickly and effectively to the aftermath of an earthquake.

> **Exam tip**
> You are more likely to show a grasp of the depth and detail of this topic if you can discuss these various examples in detail in your exam answers. Learn them well.

> **Exam tip**
> The CCEA Specification requires you to evaluate the preparations for and responses to one earthquake. You can deepen your understanding of this by researching these aspects for an earthquake which occurred in an LEDC, for example Haiti. Here, a much smaller earthquake killed over 250,000 people. Do some basic research to find out why, comparing it to the Japanese case study on p. 26–28.

Case study

Tohoku, Japan, 2011

In March 2011, a magnitude 9.0 earthquake struck off the northeast coast of Japan at Tohoku (Figure 12). The seismic waves it sent out shook thousands of Japan's buildings and caused liquefaction across hundreds of kilometres of the coastal plain nearby. But it was the devastating tsunami that was triggered by the earthquake that caused the vast majority of death and destruction.

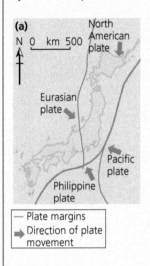

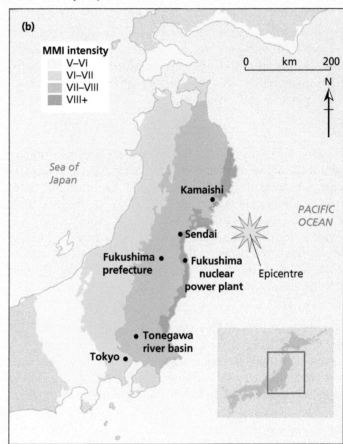

Figure 12 Map of Japan showing the Tohoku earthquake

Preparation: building codes

There are both positive and negative aspects of the preparations for the earthquake:

Positive

- Japan has the most rigorous earthquake building codes in the world. In 1950, the *kyu-taishin* earthquake building codes were first legally enforced across the entire country. Following a magnitude 7.4 quake in 1978, even stricter codes were introduced in 1981, known as *shin-taishin*.
- As a result of these codes, there was remarkably little structural damage in Japan. For example, no buildings collapsed in Tokyo. There was some damage to foundations in hilly areas and reclaimed lands in Kanto region (96 cities, towns and villages along the coast of Tokyo Bay and Tonegawa river basin) due to liquefaction. But, in total, only 18,000 of the 125,000 buildings destroyed (14%) were the result of liquefaction or seismic shaking. It is estimated that, out of a total of 19,781 dead or missing, only around 250 deaths occurred due to building collapse (1.3% of all deaths).

➜

Negative

- On the other hand, there were some issues in building construction that became clear following the quake. Even where buildings did not collapse, there was falling of elements like ceilings, even in *shin-taishin* buildings. There was some building collapse: for example, in the Fukushima Prefecture, the ground floor of some mid-rise apartment buildings collapsed. This occurred mostly in buildings constructed to the *kyu-taishin* standards; the better designed buildings tended not to experience collapse. But, worryingly, there are still around 20–30% of buildings in Japan which comply only to the older *kyu-taishin* codes.

Preparation: tsunami walls

Positive

- Prior to the 2011 earthquake, Japan had invested heavily in sea wall defences against tsunamis. Around 40% of the vulnerable coastline was protected by sea walls, some up to 12 m high. For example, the tsunamic defence wall in Kamaishi Tsunami Protection Breakwater, completed in 2009 after 30 years of construction, entered the Guinness World Records as the world's deepest breakwater.

Negative

- Despite this, 94.5% of the deaths from this quake happened due to the tsunami. The initial tsunami wave height estimates from the JMA were too low (around 3–4 m) because their initial magnitude estimates for the quake were too weak (around magnitude 7.8). This meant that many people believed the sea walls would protect them. Only when it was too late was the warning upgraded; by that time many people didn't have time to save themselves by following an evacuation route.

In conclusion, the building codes undoubtedly reduced the death totals considerably in this massive earthquake; damage due to seismic shaking and liquefaction was minimal. However, despite the investment in tsunami sea walls, flawed tsunami warnings meant that the tsunami had a devastating impact.

Response: Fukushima nuclear meltdown

There are both positive and negative aspects of the responses to the earthquake:

Positive

- The earthquake led to one of the biggest nuclear disasters of recent years when the reactors in the nuclear power plant at Fukushima overheated, causing a series of explosions which released radioactivity into the air.
- The immediate response was evacuation: initially, a 20 km evacuation zone was set up (and 80,000 people were evacuated) and 136,000 people living 20–30 km away from the plant were told to stay indoors. It took 6 months and 3,000 people working in the site before the meltdown was contained and no more radiation was being emitted.

Negative

- However, in the longer term, there remains increased risk of cancer among the population (in eastern Fukushima Prefecture, radiation levels are 15 times higher than they were before the tsunami) and impacts on local food supplies (it is thought that certain fish such as cod and sole will be inedible until at least 2021).

Response: tsunami sea walls

Positive

- Following the 2011 earthquake, the Japanese authorities have produced plans to construct a massive sea wall to improve their tsunami defences. This $6.8 billion plan will link 440 sections of wall together to form a 400 km-long (250-mile) wall that in some places will stand more than 12 metres tall.

→

Negative
- This is a controversial plan. First, there are debates about the effectiveness of tsunami sea walls. The sea wall at Fudai did protect that village. But the sea wall at Kamaishi, supposedly protected by a $1.6 billion breakwater (the world's largest), collapsed under the impact of the tsunami, and the city was devastated as a result. It is estimated that nearly 90% of sea walls along the northeast coast were similarly damaged.
- Secondly, there are wider environmental concerns (as it will disrupt marine ecosystems) and economic concerns (as it will negatively impact local fisheries).

Overall, given the scale of the 9.0 quake, the response of the Japanese authorities was good. Early rescue intervention and nuclear containment certainly saved many lives. Furthermore, significant investment has been made in hard engineering to try to protect against future tsunamis. However, the response was not perfect: issues about the efficacy of the new tsunami defences, and ongoing issues with nuclear contamination, cause concerns.

Summary

- Earthquakes occur when there is a sudden release of built-up seismic strain energy along a fault. The waves sent out from the earthquake's focus take three forms: P waves, S waves and L waves, each with different characteristics.
- Earthquake foci tend to be shallower at constructive, conservative and collision zones, as no subduction takes place here. The subduction at destructive margins causes shallow and deep earthquakes forming a pattern known as the Benioff Zone.

- Earthquakes can have a number of impacts including: seismic shaking, which can be very destructive for infrastructure; liquefaction, which tends to occur on soils that are sandy and quite saturated; and tsunamis, which can cause significant loss of life and damage to infrastructure.
- Various attempts have been made to develop theories to predict earthquakes including the seismic gap theory and dilation theory. However, these have had very little success.

■ Option B Tropical ecosystems: nature and sustainability

Locations and climates of major tropical biomes

Distribution, climatic and biomass characteristics of ecosystems

The distribution of tropical biomes

A **biome** is a large-scale global ecosystem that occupies a distinct region across the Earth (Figure 13).

An **ecosystem** is a community of plants and animals interacting with each other and the environment in which they live.

Ecosystems across the world are named after the dominant vegetation type that can be found in that place.

Tropical biomes can be found between the equator and the Tropic of Cancer (to the north at 23.5°N) and the Tropic of Capricorn (to the south at 23.5°S).

> **Exam tip**
>
> Make sure you can describe the location of the major tropical biomes from a world map.

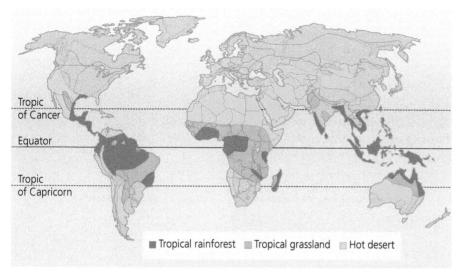

Figure 13 Map of major tropical biomes

Tropical rainforest

Tropical rainforest is usually found between 10° north and 10° south of the equator. Much of South America is covered in rainforest, including the massive 7 million square kilometres of the Amazon rainforest in Brazil. Parts of western Africa also have rainforest including the Congo rainforest. The islands and part of the Asian mainland are populated by tropical rainforest. Many consider the southeast Asian rainforests to be the oldest on the planet and they can be found in Burma, Thailand, Vietnam, Malaysia, the Philippines and Indonesia.

Tropical grassland

Tropical grassland is sometimes called savanna. This area covers a wide expanse across Africa, South America and parts of India and Australia (between 5° and 20° north and south of the equator). There are areas of grassland in India (stretching into Nepal) and across northern Australia (covering around 23% of Australia's land through the Northern Territory and Queensland). The northern areas of South America contain a tropical grassland plain called 'Los Llanos' in Colombia and Venezuela and south of the equator the south Brazilian Campos grasslands are a unique ecosystem. The largest area of tropical grassland spans from Guinea, Sierra Leone, Cote d'Ivoire, Ghana, Togo, Benin, Nigeria in the west of Africa into the Central African Republic, Cameroon, Congo and towards the eastern African countries of Sudan, South Sudan, Ethiopia, Rwanda, Burundi, Kenya, Tanzania, Malawi, Zambia and Zimbabwe.

Desert

Deserts are estimated to take up about one-third of the Earth's land surface. Hot deserts are usually found between 15° and 30° north and south of the equator. They usually stretch from the western coast inland. In North America, the Great Basin Desert covers 492,000 km^2 in California, Nevada and Utah. In Australia, the Great Victoria desert is one of many deserts that cover vast areas of Western Australia and South Australia. The Arabian desert and Syrian desert cover the vast majority of land in the Middle East. In Africa, the Kalahari and Namib deserts are located to the southwest of Africa

Knowledge check 7

What are the main latitude measurements for each of the three tropical biomes?

Exam tip

Make sure that you know the facts and details about the location of each of the tropical ecosystems.

covering 647,500 km² across Botswana, Namibia and South Africa. The Sahara is the largest hot desert, covering over 9 million km² and stretching from western Sahara, Morocco, Algeria, Libya, Niger and Chad into Egypt, Sudan, South Sudan and Ethiopia.

The climatic characteristics of tropical biomes

Climate is the most important factor that determines the key properties in each tropical biome. The main climatic characteristics in each biome include precipitation, temperature, light intensity and wind. Here we will focus on the impact that precipitation and temperature play in the tropical biomes (Figure 14).

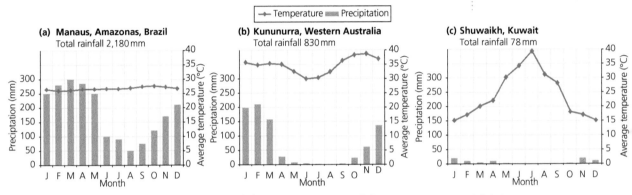

Figure 14 Climate graphs for (a) tropical rainforest, (b) tropical grassland, (c) desert

Tropical rainforest

Temperatures remain consistently high throughout the year. The sun is always high in the sky which means that the temperature range is very small. In Manaus, located in the Amazonas region to the north of Brazil, the mean temperature varies between 26°C and 27.7°C. The range only varies through the year by about 1.7°C.

The amount of annual **rainfall** usually exceeds 2,000 mm. In Manaus, the total is 2,180 mm. The quantity of rainfall is broadly the same through the year but slightly less rain falls in the summer months (June to September). Each day in the rainforest warm air will be forced to rise quickly as a result of the convergence of the trade winds at the ITCZ. This causes convection currents to push up the warm, unstable air to form towering cumulonimbus clouds that will often bring thunder and lightning in the late afternoon. In Manaus rainfall is expected on at least 265 days every year.

Tropical grassland

Temperatures in the savanna grassland also remain consistently high throughout the year. The sun is always high in the sky which means that the temperature range is very small. In Kununurra, in Western Australia, the mean temperature varies between 30.1°C and 38.9°C. The range varies through the year by about 8.8°C.

The amount of annual **rainfall** can be unpredictable. Rainy seasons can last over five months in the summer and usually there is a winter 'drought'. In Kununurra, the total is 830 mm. As Kununurra is in the southern hemisphere, the winter 'drought' is experienced from April to September. The rainy season in the southern hemisphere occurs from October through to March, with a maximum in February.

Desert

Temperatures in the desert are very high for most of the year and can peak at a very high level through the year. The sun is high in the sky through the summer months. In Shuwaikh, in Kuwait, the mean temperature varies between 15°C (in January) and 39°C (in July). There are usually 14 hours of daylight and 10 hours of darkness in the desert in the summer. The diurnal range in the desert can be wide: temperatures during the day might reach 40°C but at night the clear skies will cause a rapid cooling.

The amount of annual **rainfall** is very low in the desert. In Shuwaikh, the typical total is only 78 mm. An area will be classed as a desert if the annual total falls below 250 mm. In the northern hemisphere it is usual for any rain to occur through the summer months.

The biomass characteristics of tropical biomes

Biomass is an ecological term to show the amount of living organic material that can be found in an area or ecosystem. It is made up of animal life (fauna) and plant life (flora) and is usually expressed as the dry weight of matter per unit area or the total mass in the community.

All organisms need energy for growth and reproduction, but there is a limited amount of energy available within any ecosystem.

The rate at which any biome can produce material is known as productivity. The total live biomass on the Earth is around 560 billion tonnes of carbon with a total annual production of biomass of just over 100 million tonnes of carbon per year.

The level of **primary productivity** (Table 3) in an ecosystem is the rate at which producers will build the amount of biomass. This will set the amount of energy available throughout the trophic levels in the ecosystem.

Table 3 Productivity in the tropical biomes

	Average net primary productivity $(g\,m^{-2}y^{-1})$	Total production biomass (billion tonnes of carbon per year)	Total biomass (billion tonnes)
Tropical rainforest	2200	16	765
Tropical grassland	950	9	60
Desert	40	0.15	13

The global distribution, location and nature of biomass

Figure 15 shows the global distribution of biomass across the world. Forests cover over 4 billion hectares globally, or over 30% of the Earth's land area. They also account for over 75% of all primary production. The tropical rainforests of South America, Africa and southeast Asia make up a huge amount of biomass. There is less available in the tropical grassland and very little vegetation present in the desert regions.

Knowledge check 8

What are the average annual amounts of rainfall in each of the tropical biomes?

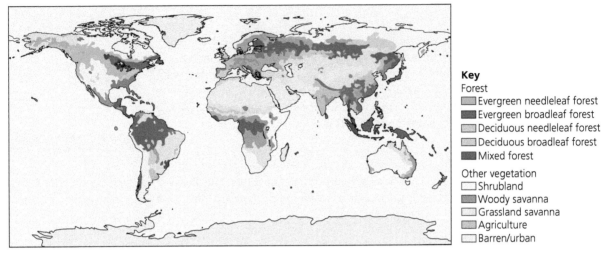

Figure 15 Global distribution of biomass in 2011

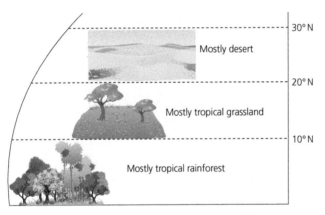

Figure 16 Changes to biomass in tropical areas

Tropical rainforest

The high amounts of rainfall and high temperatures allow for a rich diversity of vegetation to grow and survive in tropical rainforests. The amount of vegetation that can be supported will be reduced steadily until the impact of the sinking air at 30°N and 30°S is noted in the desert areas.

There is a huge amount of vegetation within the tropical rainforest — over 40% of all global vegetation can be found here. There is much biodiversity within the rainforest. Around 75% of all the biotic species are found here. Two-thirds of all flowering plants can be found in rainforests and a single hectare of rainforest may contain over 42,000 species of insect. The high temperatures, high amounts of solar radiation and heavy rainfall create balmy conditions where there is a high rate of decay and recycling of nutrients. Most of the trees, such as mahogany, ebony, palm, brazil nut and rubber, are hardwoods and are evergreen in appearance. Tall emergent trees can reach up to 50 m in height as they aim to break through the various layers of canopy so that they can access maximum amounts of sunlight. Many of the trees are covered in epiphytes which can use the tree for support, and lianas.

Exam tip

You should be aware of the differences in the amount of biomass across each of the three ecosystems and be able to show how this can change.

Exam tip

Use Figure 16 to understand that the distribution of biomass is located across particular latitudes.

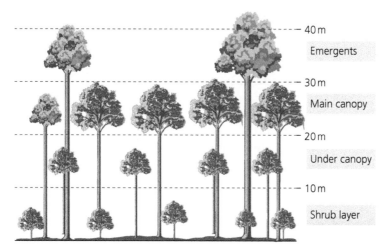

Figure 17 Vegetation layers in tropical rainforest

The main and under canopy levels provide a near continuous vegetation cover that lets relatively little light through to the layers beneath (Figure 17). Many of the tallest trees have developed buttress roots that will support them. The trunks of the trees are usually branchless in the under canopy as there is no light for photosynthesis and most of the energy of the tree is focused on pushing the tree crown towards the sunlight.

There is also a rich diversity of animal life within the rainforest: nearly 600 species of mammal, 2,000 types of bird and 2,000 species of amphibians and fish, plus countless numbers of insects and micro-organisms. Some of the larger mammals such as gorilla, chimpanzee, orang-utan and jaguar are endangered species. Many of the animals and birds have adapted to live in this area. Sloths, monkeys and anteaters have adapted to be able to live in trees, while others live on the ground and exist by collecting seeds and fruit that fall from the canopy layers.

Tropical grassland

Vegetation in the tropical grassland can be quite varied. Rainfall amounts can still be relatively high, although they will be very seasonal. Many tropical grassland areas have enough rainfall to support deciduous and thorn-scrub forest vegetation. The grasses in this area can grow up to 3.5 m high and they will have long root systems that allow them to reach water stored deep underground. Many of the trees and plants have adapted some drought-resisting features, e.g. acacia has a deep and complicated root system, whereas baobab trees store water in their trunks to help them survive the dry season. Both the grasses and trees are deciduous and will lose their leaves in the dry season.

Many of the world's largest animals are found in the tropical grasslands as they can graze on the vegetation easily, e.g. elephants, zebras, antelopes and wildebeest. The amount of biodiversity is lower than in the rainforest, e.g. in Tanzania there are 375 species of mammal, 1,000 species of bird and over 800 species of reptiles, amphibians and fish. The grassland also supports a huge number of burrowing rodent animals. Large carnivores, e.g. lions and hyena, will feed on both the rodent and herbivore animals. One of the major features of the grassland is the great

Exam tip

As well as knowing the differences in the layers within the tropical rainforest, you might like to look at some of the different ways in which plants and animals have adapted so they can survive and thrive in the tropical rainforest.

annual migrations that can take place as animals are forced to look for water during the dry seasons.

Desert

The amount of biomass within the desert is significantly lower than in the other biomes. Desert vegetation is usually sparse and scattered as there is not enough rainfall and moisture to support a continuous spread of plant cover. Individual plants have developed a number of drought-resistant features that can enable their survival, such as seeds that will germinate quickly when rain falls. Others have thick waxy leaves that can be shed in drought conditions or they will be able to store water within their stems, like cacti. Other shrubs might have very long roots that can delve deep into the ground to find any water available.

There are relatively few mammals that are able to survive the hot daily temperatures or the daily diurnal range. Most will hide during the day and are largely nocturnal. Insects and reptiles are more common in the desert and they will have waterproof skins.

The Hadley cell and ITCZ

The Hadley cell and ITCZ control location and climate in tropical forest, tropical grassland and desert ecosystems.

The Hadley cell

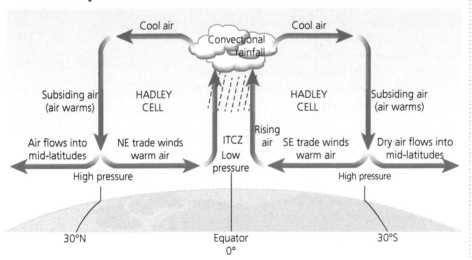

Figure 18 The Hadley cell

The **Hadley cell** (Figure 18) is part of the tri-cellular model that controls the global circulation of air. The Earth's atmospheric engine gets its energy from the direct solar heating at the equator. Heat is transferred to the air above and the air rises and cools. Rising air means that low pressure, known as the inter-tropical convergence zone (**ITCZ**), is left at the surface. The rising air diverges and flows towards the poles (both north and south) and sinks again around 30° north and south of the equator. This warms and produces high pressure. The air is then diverted back towards the equator at a much lower level while some air moves towards the mid-latitudes to redress the global heat imbalance.

> **Exam tip**
>
> There is a lot of technical detail required to answer questions about the Hadley cell and the role and movement of the ITCZ. Learn this carefully.

> **Knowledge check 9**
>
> Describe how the Hadley cell causes high pressure at the tropics.

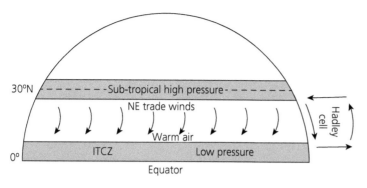

Figure 19 Tropical air circulation

The northeast and southeast trade winds blow from the areas of sub-tropical high pressure and meet around the equator. The winds converge at this point and the air is heated by solar radiation, causing the air to rise, condense and form clouds (Figure 19).

Convectional rainfall will bring a large amount of heavy precipitation and thunderstorms. Rainfall occurs on a daily basis with convectional cells bringing heavy rain storms most afternoons. Each day the ITCZ allows the warm air to rise, cool, condense and for clouds to form quickly. The intense heat in the area will fuel these processes to happen at a very fast rate and towering cumulonimbus clouds will be formed. Air pressure will remain low throughout the year.

As the air moves north and south of the equator as part of the Hadley cell, the air starts to descend between 20° and 30° north and south of the equator. Sinking air produces an area of high pressure at the surface where there are clear skies and weather conditions are stable and there is little or no precipitation, and hot desert climates will be found here. This sinking air also causes the sub-tropical high pressure belt which creates the trade winds and forces the air to move back towards the equator at ground level. Most deserts, including the Sahara and Arabian deserts, are found on the western edge of continents due to the impact that cold ocean currents have on the land mass.

The ITCZ

The Hadley cell is moved due to a seasonal shift which moves the position of the ITCZ. It is the movement of this ITCZ which has a huge impact on the climate characteristics of the tropical grassland (Figure 20).

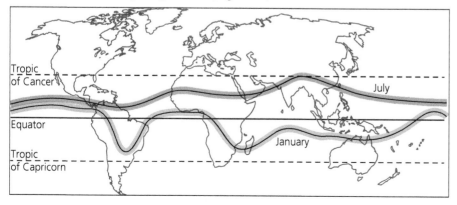

Figure 20 Movement of the ITCZ

In the winter (January) the ITCZ will move south leaving some of the northern hemisphere tropical grassland areas under the high pressure area, and they will not receive much rain (Figure 21). This becomes the dry season (for example in the south of Brazil and in the savanna grasslands in the east of Africa).

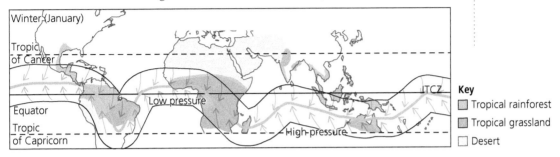

Figure 21 Impact of ITCZ movement in January

In the summer (July) the ITCZ moves north which brings the low pressure over the tropical grasslands in the northern hemisphere. With the low pressure now overhead in some places there will be some heavy rainfall. As the ITCZ is pulled north, the area of high pressure will also be pulled north which allows more rainfall due to the impact of the southwesterlies (in the northern hemisphere) and the northwesterlies (in the southern hemisphere) (Figure 22).

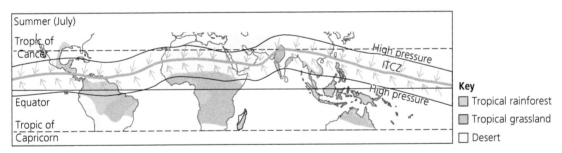

Figure 22 Impact of ITCZ movement in July

Summary

- The distribution of tropical rainforests is largely on the equator. As you go further north or south you will come to tropical grassland and then at the tropics you will find a desert ecosystem.
- The climate characteristics of the tropical rainforest are hot and very wet. The tropical grassland is hot and damp and the desert is very hot and very dry.
- Biomass is an ecological term used to describe the amount of living organic material that can be found in an area or ecosystem. It is made up of animal (fauna) and plant life (flora).
- The Hadley cell and the ITCZ are important elements in transferring global heat energy and producing low and high pressure points around the world.

Management and sustainability in arid/semi-arid tropical ecosystems

Arid and semi-arid tropical ecosystems (deserts) are usually found between 15° and 30° north and south of the equator. They are due to falling air which creates high pressure weather systems at the surface. The descending air means that there is very little precipitation in these places causing the semi-arid and arid conditions to develop.

Sustainable development is described as being socioeconomic progress that meets the needs of today's population without harming the ability of future generations to meet their needs. A balance needs to be reached between the environmental and socioeconomic needs within the area so that people can survive and make money but not destroy the ecosystem at the same time.

Knowledge check 10

What is sustainable development?

Use of irrigation in arid/semi-arid tropical environments

Arid and semi-arid climates are characterised by a lack of water. Irrigation is the artificial supply of water to the land. This water might come from a redirection of water from another area, river, lake or an underground source. Irrigation will usually be carried out to allow the cultivation of the land.

Irrigation has become a very important part of the modernisation of agriculture in recent years across the world. It was of vital importance to Asia's Green Revolution between the 1930s and 1960s. Irrigation across African countries currently involves only around 5.4% of all cultivated land. This compares to the global average of 20% and 40% across Asia. As the global population continues to increase, there is a need for many countries to consider more irrigation strategies so that food production can be increased further.

There are three types of irrigation:

- **Gravity irrigation**: this uses a series of channels and ditches to move water by gravity. These flood or basin irrigation systems usually divert water from a river through canals and flood the whole field. These can help create rice paddies.
- **Spray irrigation**: this delivers water to a large area in a way that sprays or sprinkles water to mimic rainfall. Water can be lifted to the surface through a variety of human- or animal-powered technologies or more modern pump systems.
- **Drip irrigation**: this delivers water via pipes to individual plants. Holes in the pipe are placed next to the plants and the holes deliver water directly to the plant. This process will help to reduce water loss by evaporation and leakage.

Environmental and socioeconomic benefits and problems of irrigation

In the past, basic channels and aqueducts were used to spread water quickly from one place to another. Modern irrigation methods are more complicated — electric pumps will suck water up from great depths and then metal pipes can carry the water long distances before it is released onto the land.

Exam tip

There is quite a long list of environmental impacts of irrigation, both positive and negative, but also consider whether you think the positives outweigh the negatives.

Environmental impacts

Table 4 Environmental impacts of irrigation

Environmental benefits of irrigation	Environmental problems of irrigation
Irrigation systems can be very simple and bring water onto dry land quickly and effectively. Crops will be able to survive in drought conditions, e.g. cotton or rice.	Irrigation can cause reductions in groundwater, rivers, lakes and overland flow. There can be increased evaporation in the irrigated area and groundwater will take a much longer time to recharge to 'normal' levels.
Excess water infiltrates into the soil recharging the soil moisture and it will percolate into the rocks to recharge the groundwater stores (and raise the water table).	Waterlogging: irrigation could lead to the soil becoming saturated to a point where this is damaging to the agricultural activity. Most crops need oxygen and waterlogging can stop air getting into the soil, which can cause soil erosion.
Spray irrigation systems are more easily controlled than gravity flow systems. Sprinklers can be turned off and on as necessary. The sprinkler system can also be easily directed to the area of most need — causing less wastage.	Soil salinisation: in areas with high amounts of evaporation, water is drawn back up towards the surface (by capillary action) and salts that are usually kept deep within the soil will be brought up to the surface.
Drip irrigation systems are the most efficient. The water can be controlled carefully and can react to the demand of the particular crops. Only the plants that need water will receive it. There will also be a lot less water lost through evaporation (and even less if the irrigation system is installed beneath the surface).	Ecological damage: excessive use of irrigation measures can cause a breakdown in the natural ecosystems in an area. If groundwater is used excessively this causes the water table to fall. In Texas, USA, the water table has in some places fallen by up to 50 m.
There is evidence to suggest that irrigation can increase the amount of precipitation in an area. Increased evaporation and transpiration will lead to more moisture in the atmosphere.	Reduced river flow: a lot of irrigation water is taken from rivers. This can mean that ecologically important wetlands might be lost.
Improved food supply means that diseases in vegetation and in animals can be reduced causing fewer environmental issues in an area.	Stagnant water caused by irrigation on the surface of the soil can also cause increases in the incidence of water-borne diseases like malaria, yellow fever, dengue and bilharzia.

Socioeconomic impacts

Table 5 Socioeconomic impacts of irrigation

Socioeconomic benefits of irrigation	Socioeconomic problems of irrigation
The technology involved in many aspect of irrigation is simple and can be developed anywhere.	Sprinkler and drip systems can be expensive which means that farmers in LEDCs might not be able to afford these systems.
Some technology can be inexpensive, using a lot of manpower.	Sprinkler systems require a central source of water and the system will require a high amount of pressure to pump the water through the pipe system. They might also need access to electricity to power the pump system.
The use of irrigation measures will increase the amount of land that can be used for farming. It increases productivity, increasing the size of yields and harvests.	Waterlogging and salinisation of the soil can both lead to the soil becoming less useful for agriculture. Irrigation requires careful management to ensure that the correct amount of water is used on the land.
Improved food supply and food security in a country also reduces malnutrition and reduces the country's dependency on food imports. Excess food can be exported — helping with trade balances and the overall economy of the country.	Salinisation can lead to soil erosion which in turn can lead to desertification.

Possible solutions to the problems of irrigation in arid/semi-arid tropical environments

Irrigation projects will become increasingly complicated and technologically advanced in the future. About 40% of the world's total food supply comes from irrigated land.

Irrigation techniques can 'lose' water in a number of ways. In each case, the irrigation techniques need to be improved and made more efficient. Water can seep out of reservoirs or be lost through leaks in canals and pipes before it even gets to the field. New technologies have been developed to reduce irrigation losses. For example, precision land-levelling uses laser-guided equipment to level a field so that water will flow uniformly through the soil and will not collect in gullies or ditches across the field. Increased use of drip irrigation and sprinkler methods also enhance water delivery and ensure that water reaches the places where it is most needed and that less water is lost to deep percolation or surface run-off from the field. In some places unmanned autonomous systems (UAS) or 'drones' can be used to monitor agricultural systems and to provide imagery that helps a more precise use of irrigation systems. However, many of the new technological solutions to irrigation issues are extremely expensive and are only applicable in the richer MEDCs.

Knowledge check 11

Describe how waterlogging and salinisation can impact crop harvests.

Exam tip

You should use the general points about the benefits and problems of irrigation (Tables 4 and 5) as the starting point for your case study about Pakistan and then use the information below for more precision.

Case study

The management of irrigation along the Indus river, Pakistan

Use of irrigation

The vast majority of Pakistan is classified as either a warm desert or warm semi-arid climate. This means that drought can be frequent. The drought of 1998–2002 was considered to be one of the worst in over 50 years. Nearly 200 million people live in Pakistan and there is much competition for resources and water in particular.

The growth and prosperity of the population in this area is linked to the growth of irrigation techniques, drainage, sewers, canals and barrages.

The north of Pakistan is found in the foothills of the Himalayas where there is much more rainfall. (The average precipitation in Islamabad is 1,250mm each year which contrasts with 230mm each year in Hyderabad.) This rain feeds the Indus river and the river moves the water through the country and through the drier desert regions in the south, providing an opportunity for irrigation which will allow cultivation of an increased amount of farmland.

The Indus Basin Irrigation System (IBIS) (Figure 23) is made up of three major reservoirs, 16 barrages,

two head-works, two siphons, 12 inter-river link canals, 44 canal systems and more than 107,000 water courses. The total length of all the canals is 56,073km and the total amount of water courses, channels and ditches amounts to around 1.6 million km. The IBIS is the largest irrigation system in the world, irrigating over 18 million hectares of farm land and allowing for the production of wheat, rice, fruits, vegetables, sugarcane, maize and cotton.

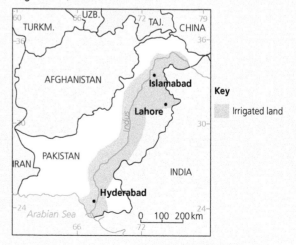

Figure 23 Map of irrigated land in Pakistan

Environmental and socioeconomic benefits and problems of irrigation

There are three main sources of water in the Indus Basin:

1 river water in the Indus river
2 rain water
3 groundwater (around 40% of crop water in Pakistan)

Agriculture is an important part of the Pakistan economy. As well as providing food security for the country, agriculture also makes up 23% of the GDP, 70% of total export earnings and employs 54% of the labour force.

One of the biggest problems in the Indus valley concerns the amount of water that individual farmers are using. Farmers in Pakistan get their share of irrigation waters on a rotational basis. This means that farmers often use more water than they require to ensure that they retain their share of water. There is much wastage of irrigation water.

Salinisation has become a big problem in Pakistan. There is a large amount of natural salts stored in the soil and rocks. High temperatures will cause these salts to be brought to the surface and this can lead to infertility and soil erosion.

Irrigation measures including dams, barrages and river diversion have meant that parts of the river do not flood. This deprives the floodplains of alluvium and silt deposits that supply nutrients to crops and allow them to grow without the need to buy artificial fertiliser.

Waterlogging can also become a problem. Farmers will traditionally over-use the irrigation water which means that the soil will often be saturated. This can lead to water-borne diseases (like bilharzia). This can also cause fast surface run-off in flood conditions — for example, when the Indus river undergoes flood conditions (especially following monsoon rainfall in July/August), the damage done due to flooding will be enhanced.

Possible solutions to the problems

In order to manage some of the problems created by the different irrigation practices in Pakistan some new solutions have been suggested.

■ Changing crops: some farmers have moved from wheat to rice farming as the rice plants can tolerate the salty and waterlogged conditions more readily. Farmers have also been encouraged to plant more trees, in particular fruit trees which will soak up and store water more readily. In areas where salinisation had really taken hold, plants that are more tolerant to the saline conditions (halophytes) could be planted instead of the traditional crops. These might include various types of grass (for pastoral farming) and fruit trees such as palm and coconut.

■ Changing practices: farmers have also been encouraged to leave some land fallow for some time to allow the rain to wash salts back through the surface of the soil.

■ Modern irrigation systems: farmers have also been encouraged to sink their own tube wells which could be used for their own needs. This would also help to lower the level of the water table and prevent flooding. Farmers have been encouraged to use more modern irrigation techniques such as sprinkler and drip irrigation systems that would be more effective in delivering water directly to the plants that need it. The solar-powered irrigation system (SPIS) has been launched recently which introduces a solar powered pumping system with a drip-based and sprinkler-based system. The SPIS has minimal operating costs and has been manufactured using sustainable materials. The impact is that water is used more efficiently and more land can be brought into effective cultivation.

Summary

- Arid/semi-arid tropical ecosystems (deserts) are found between 15° and 30° north and south of the equator and are caused by falling air as part of the Hadley cell.
- Irrigation is the artificial supply of water to the land. The main types of irrigation are gravity, spray and drip irrigation.
- Irrigation can bring waterlogging or salinisation to the soil which can reduce yields and harvests and can lead to soil erosion and eventually desertification.
- Modern irrigation methods are much more accurate in the amount of water that is used on the soil.
- The Indus river in Pakistan is the central feature of one of the world's biggest and most successful irrigation schemes. Water is moved from the wet north to the dry south of the country using the river and a series of canals and drainage ditches.

Management and sustainability in the tropical forest environment

Evaluation: large-scale development

Large-scale development poses a threat to the trophic structure, nutrient cycle and zonal soil of the tropical forest ecosystem.

Trophic structure

Energy is transferred up through trophic levels (Figure 24). The first trophic level consists of the plants. As they produce their own food directly, they are called producers (or autotrophs). In the second trophic level, the herbivores (plant eaters) consume plants from the first level. In turn, these herbivores provide food for the third trophic level, the smaller carnivores (meat eaters). The fourth trophic level is made up of the larger carnivores and omnivores (plant and meat eaters).

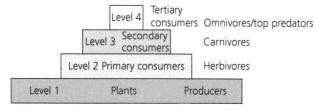

Figure 24 The trophic pyramid

Nutrient cycling

Nutrients mainly cycle between stores within the ecosystem (Figure 25). The nutrients will leave the biomass store and are transferred by fallout into the litter store. The dead organic matter is broken down by the decomposers and transferred into the soil store. Here, nutrients are available to plants via their roots and are transferred by uptake into the biomass store and the cycle continues. In addition, there are two inputs of nutrients, such as carbon dissolved in precipitation or minerals weathered from the bedrock, and two outputs (losses via run-off from the litter store or leaching from the soil store).

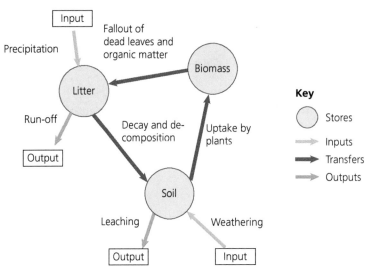

Figure 25 The Gersmehl model of nutrient cycling

Zonal soils

A zonal soil occurs when the nature of the soil reflects the climatic aspects of a world region. Usually soils will differ depending on local conditions, but a zonal soil shows the total control that climate has in a particular area. In the tropical rainforest the zonal soil has developed in response to the extremely hot temperatures, high humidity (over 88%) and very wet conditions. The main soils found across most tropical rainforests are controlled by the acidity, low amounts of nutrients and poor drainage. The soils are described as being oxisols (or sometimes latosol/ferralsol).

Threat of large-scale development

There is a need for additional areas of agricultural land and many of the low-latitude rainforests have been threatened with clearing, logging and cultivation for cash crops and animal grazing.

In the past these areas of tropical rainforest were farmed by native people using slash-and-burn methods where they would cut down all the vegetation across a small area. Burning the vegetation on the site would release the trapped nutrients and return some of them to the soil, making a small amount of the nutrients available for crop growth. Over time, the harvests will be depleted as the nutrients are used up. After a few years, the old field will be abandoned and the native people will move on to repeat the process in another untouched area. Eventually, the rainforest will reclaim the cleared area.

In more recent times a more modern, industrial and intensive approach has been taken to agriculture in the rainforest. Large areas are cleared and when these too lose their usefulness, the areas are too large for seeds and native plant species to re-populate the cleared area.

The expanse of the tropical rainforest defies the fragility of the forest. The delicate ecosystem of the rainforest is a very fragile environment and it does not take much for the ecosystem to break down.

The Amazon basin

The Amazon rainforest covers an area of 7 million square kilometres. Most of the forest is found in Brazil (60%), but it also extends into Peru (13%), Colombia (10%) and smaller amounts in Venezuela, Ecuador, Bolivia, Guyana, Suriname and French Guiana (Figure 26).

Some scientists think that the Amazon is now equivalent to half of the planet's remaining rainforests, with an estimated 390 billion trees across 16,000 species. The biodiversity of the Amazon is unequalled globally. One in ten of the known species in the world lives in the rainforest — an estimated 2.5 million insect species, 40,000 plant species, 2,200 fish, 1,300 birds, 427 mammals, 428 amphibians and 378 reptiles.

Figure 26 The Amazon drainage basin

In recent years there has been a huge increase in the amount of deforestation taking place across the Amazon basin. The building of the Trans-Amazonian highway in the 1970s opened up parts of the rainforest that had previously been inaccessible. This led to an increasing amount of the rainforest being lost to industrial pastoral and cattle farming.

Impact on trophic structure

The hot and wet growing conditions create a perfect environment for plants to grow quickly. The vegetation in the rainforest uses the process of photosynthesis to convert sunlight into plant material. This produces huge amounts of energy that the primary consumers will start to feed on. Plant-eating organisms, insects, birds, frogs, parrots and mammals will be found at different levels within the rainforest ecosystem. Herbivores will in turn be eaten by the carnivores found at the secondary consumer level. Larger animals like bats and some types of snake will consume the smaller animals. The top of the trophic pyramid is reserved for the tertiary consumers such as the larger snakes and big cats like the jaguar and puma. These carnivores act as the main predators in the ecosystem (Figure 27).

The hot and wet climatic conditions often mean that vegetation and animal matter will be broken down (decompose) very quickly after they die. A variety of insects such as ants and termites work alongside bacteria, fungi and other micro-organisms to break down dead material quickly.

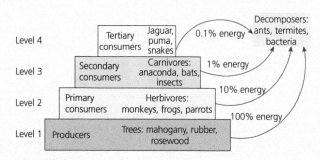

Figure 27 Trophic pyramid for the tropical rainforest

Commercial logging and farming in the Amazon have a huge impact on the amount of vegetation available in an area. The tall emergent trees which make up the climax vegetation in the ecosystem would not be present and as a result the number of primary producers would be absent from the bottom of the trophic levels. This would have a knock-on effect ➡

through each of the trophic levels — the primary carnivores would not have enough food/energy available to them and as a result many would starve and numbers would be reduced. The secondary consumers at the third and fourth trophic levels would equally have limited access to food supplies and their energy/food needs would not be satisfied and many would disappear from the ecosystem. The relationships within the rainforest are often incredibly delicate — to the point that some animals might only be able to eat the leaves of one tree in an area.

Impact on nutrient cycling

In most other nutrient cycling models for global biomes there is a balance between the size of the biomass, soil and litter involved. However, in the tropical rainforest there is a much bigger amount of biomass present within the ecosystem (Figure 28). The vast majority (over 90%) of the nutrients will remain locked up within the biomass. Litter will be falling continuously where it will decay and decompose very quickly due to the climatic conditions. The soil does not retain huge amounts of nutrients. Any nutrients will be taken up by the vegetation and in many cases the decaying leaf litter will not even reach the forest floor. The soil rarely stores nutrients but the uptake of nutrients back into the vegetation/ biomass will take place almost immediately. The tropical rainforest nutrient cycling system is highly effective at moving nutrients with relatively little loss through run-off or through direct leaching through the soil (Figure 29).

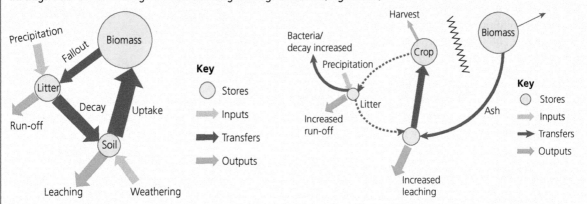

Figure 28 The Gersmehl model of nutrient cycling applied to the tropical rainforest

Figure 29 The impact of farming/logging on the Amazon basin nutrient cycle

Although the Amazon basin boasts some of the most luxuriant vegetation on the planet, these exist on some of the least fertile soils. Once the natural vegetation/biomass is removed, the amount of nutrients in the ecosystem quickly disappears. Most of the nutrient flow is through the decay and decomposition of the litter material and little nutrient content is held within the soil. This often means that the soils will very quickly become infertile. A small amount of nutrients can be re-introduced to the soil if the vegetation is burnt and ash is allowed to settle into the soil. However, any valuable nutrients added in this way will often be washed or leached quickly through the soil.

Impact on zonal soils

Oxisols found in the tropical rainforest (such as in the Amazon) are very old and are deep soils that have formed over a long period of time. Soils can be up to 20 m in depth though there can be big variations in this. The soils are relatively infertile and there is little in the way of nutrient cycling through the soil. The soils do not provide tropical plants with a rich source of nutrients such as nitrogen, phosphorus, potassium or aluminium. The soils are red in colour due to the high concentrations of iron found within the soils. Weathering is another feature of the oxisol — there is much weathering of the bedrock beneath. Leaching is common in this area — this means that as water infiltrates through the soils, it will remove humus in solution and flush any nutrients down through the soil. This leaves the upper layer of the soil increasingly acidic and deficient in minerals — often to the point that the soil has no value for any form of agricultural activity.

Any large-scale development within the Amazon basin will have catastrophic consequences for the soil. Once the soil has become exhausted of its nutrients, farming companies will often move to clear other areas of the forest. The remaining soil finds it difficult to support the regeneration of the natural vegetation and secondary species might overtake the land. Increased amounts of flooding or drought conditions can be common. Some scientists have even suggested that the rainforest is reaching a 'tipping point' in that if the drought conditions continue, the forest could irreversibly start to die. Over the last 10 years there has been an increased number of droughts within

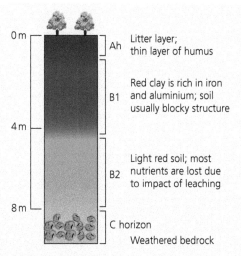

Figure 30 Soil profile for tropical rainforest

the Amazon basin. The end result is that there is an increased amount of soil erosion of the unprotected soil which means that any chance of recovery is lost forever. Industrial agriculture can also introduce a number of chemicals to the soil such as fertilisers, herbicides and pesticides which can all have negative impacts on the local ecosystem.

Negatives of large-scale development

One of the biggest commercial agriculture practices in the Amazon rainforest has been the rapid increase in soy production. In a short space of time Brazil has quickly become the leading producer of soybean (with annual production at 103 million tonnes a year). Much of this growth has been due to expansion into cleared areas of the tropical rainforest. In addition, infrastructure such as roads and ports has to be developed to move the produce. Since 1990 the area of land planted with soybeans in the Amazon has expanded at a rate of 14% per year and now covers over 8 million hectares. Recently, changes to agricultural techniques have led to double cropping through the growing year which introduces even more chemicals into the rainforest ecosystem.

> **Exam tip**
>
> There is a lot of detail in relation to this regional case study as this is a common essay question where detailed facts and figures are expected.

Table 6 gives positive and negative examples of large-scale development in the Amazon rainforest.

Table 6 Examples of large-scale development in the Amazon rainforest

Activity	Negatives	Positives
Commercial agriculture	■ An increasing amount of land is being cleared as foreign-based businesses are developing large areas of land to an industrial scale. ■ The main cash crops in Brazil are coffee, corn, rice, soybean, tobacco and sugarcane. ■ Another growth area is cattle grazing — over 250 million head of cattle are thought to be living in the Amazon basin. ■ Fertilisers, pesticides and herbicides all impact the environment.	■ Big agriculture allows local people to earn much better wages. ■ New agricultural practices (such as organic farming) have opened up new markets and created more opportunities within Brazil, while causing less environmental damage. ■ Cargill, a global agricultural grain company, has been working in Brazil for 50 years and has over 8,000 employees.

➡

Table 6 *continued*

Activity	Negatives	Positives
Logging	■ Timber sourced from the tropical rainforest is often prized. However, loggers do not promote the re-establishment of trees to replace those that have been displaced. ■ In 2014, Greenpeace noted 'the Amazon's silent crisis' — where timber laundering was noticed to have caused the illegal harvesting of rainforest lands.	■ Some species have been protected — the Brazil nut tree, native Brazilwood and Brazilian rosewood are no longer allowed to be harvested. ■ Increasingly, harvesting requires a permit and a formal management plan.
Transport	■ As new transport links have been developed (for the recent World Cup and Olympic and Paralympic events), this has opened up access to areas that were previously remote. ■ The building of the Trans-Amazonian highway (BR-230) allowed access to remote parts of the rainforest. This indirectly caused a great deal of deforestation.	■ Good transport links have allowed better access to the towns and cities in central Brazil. This has raised standards of living. ■ The roads were originally developed to allow new settlers to develop small-scale agriculture. The government motto 'land without men for men without land' helped to describe the way that agriculture was planned. ■ Many conservation bodies have been developed along the highway to monitor and support conservation work within the rainforest.
Hydroelectric power (HEP)	■ New HEP dams will flood huge areas of land. ■ New dams such as the Belo Monte dam on the Xingu river are seen as causing huge environmental issues. When the dam is completed in 2019 this will convert flooded vegetation in the huge reservoir into methane gas (a greenhouse gas).	■ The increased number of people living in Brazil, and increased need for electricity, has put huge demands on power supplies. HEP was often seen as being a simple, effective and cheap method of increasing energy capacity. ■ Dams will be able to help control the flow of water through the Amazon river and should be able to help to control flooding.
Mining	■ Huge tracts of land are often cleared to allow mining and quarrying operations to reach any valuable rocks and minerals. Brazil is the fifth largest producer of gold and also has mines for iron, tin, copper and aluminium.	■ New advancements in mining and quarrying technology allow much easier access to any rocks/minerals that are found beneath the forest canopy.

Positives of large-scale development

The reality is that there are very few positives evident when large-scale developments start within the Amazon rainforest. In most cases, human involvement brings an imbalance to the fragile ecosystem and often the rainforest is unable to recover and the forest canopy fails to be regenerated. This often leads to further problems with flash flooding and soil erosion. Every human intervention into the ecosystem has a dramatic and negative consequence. The only positives that developments bring in the Amazon are economically for the people who can tame the forest for their own financial gain.

Evaluation: sustainable development

Exam tip

The Monteverde Cloud Forest case study is a small-scale case study so details will be much more local than the wider regional case study of the Amazon. Make sure that you use the facts and figures with precision.

Across the world, people have been trying to achieve a balance between the **use** of a resource and the **appropriate management** of a resource so that it is not abused to extinction. Over the last 50 years, the tropical forest ecosystem has come under threat from many different places. Deforestation is common for timber, mining and farming. However, pressure has grown on countries to ensure that these fragile and important ecosystems are protected and preserved.

Deforestation in the Amazon rainforest has been estimated at an annual loss rate of 27,423 km² each year. Although some forest is lost due to traditional subsistence, such as nomadic farming methods, more is lost due to commercial logging, mining and mineral extraction. In recent years over 80% of deforestation has been caused due to cattle farming for the international beef and leather industries.

Case study

The Monteverde Cloud Forest, Costa Rica

Costa Rica is a small country in Central America that has one of the most diverse ecosystems in the world. The region of Monteverde (Figure 31) has over 3,000 species of plant (including orchids) and 755 different species of tree. There are 100 species of mammal, 400 bird species and 120 reptiles and amphibians. Monteverde was made into a national reserve in 1972 and is a private, non-profit reserve run by the Tropical Science Center. It contains an area of over 10,500 hectares of cloud forest and usually is visited by around 70,000 people a year. The reserve is famous for its population of quetzal, an endangered bird species that has become an icon of the regional culture and conservation efforts.

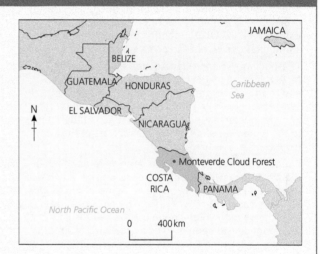

Figure 31 Location of the Monteverde Cloud Forest in Costa Rica

One way that is used to promote sustainable development in tropical rainforest areas is ecotourism. The International Ecotourism Society defines ecotourism as 'responsible travel to natural areas that conserves the environment and improves the well-being of local people by creating an international network of individuals, institutions and the tourism industry and by educating tourists and tourism professionals'.

Knowledge check 13

What is the definition of ecotourism?

➔

Evaluation of environmental development

Table 7 Evaluation of environmental development in the Monteverde Cloud Forest

Environmental positives	Environmental negatives
The development of the reserve protects the area from logging, farming and deforestation. This protects the biodiversity of the area but still allows the local people to live in a sustainable manner.	Sometimes reserves can be over-managed by humans so that natural changes to the ecosystem do not happen and the ecosystem becomes stagnant and starts to decline.
Only 2% of the forest is accessible to visitors. The rest of the forest is part of an 'absolute protection zone' where the flora and fauna is allowed to grow naturally.	Restricting access to a large area like this can be difficult to manage. Protecting the areas can be difficult — especially if there are not enough staff.
There is an extensive 'Control and Vigilance' programme that controls the natural resources in the reserve and prevents illegal hunting, deforestation, invasion and extraction of products.	The population of the area has started to grow — especially on the edges of the cloud forest and in the local town of Monteverde. This puts increasing pressure on the local ecosystem.
Research programmes have been set up to help the effective management of the area. Some long-term projects — 'Permanent Monitoring Plots' — were established in 2007 to monitor the impact of any small changes within the ecosystems.	The increase in ecotourism, while bringing much needed cash into the area, also puts pressure on the local ecosystem. Over 250,000 people visited Monteverde in 2015, which required the building of additional accommodation, facilities and transport.
The Tropical Science Center manages the cloud forest. It is a non-governmental scientific and environmental organisation that was founded in 1962. Its mission is 'the acquisition and application of knowledge pertaining to man's lasting relationship to the biological and physical resources of the tropics'.	More tourists mean that more electricity needs to be produced from sustainable sources. Getting rid of waste from hotels has also become an issue.
A number of eco-lodges have been developed inside the cloud forest. Some of these were originally made from lodges left over by loggers. Some of the lodges are very basic but others offer a more luxurious experience in the high canopy of the rainforest. The lodges have been built to high environmental standards (gaining a Certificate for Sustainable Tourism).	Even the most 'eco-friendly' tourist lodges will have a negative environmental footprint. Increases in visitor numbers have led to more trails and tracks being developed through the forest and the building of hanging bridges and 'high ropes' courses through the upper canopy of the cloud forest.

Evaluation of socioeconomic development

Table 8 Evaluation of socioeconomic development in the Monte Verde Cloud Forest

Socioeconomic positives	Socioeconomic negatives
Ecotourism has brought a very lucrative source of income to the people living in the area. The Tropical Science Center is funded through the tourism activities. In Costa Rica ecotourism brings in around $2 billion a year.	Originally only small numbers of visitors would visit the cloud forest. However, there has recently been an influx of visitors which can put more pressure on the services provided.
Jobs: the ecotourism project has brought visitors to the forest. This means that people have jobs providing services for the visitors as guides, and in shops, restaurants and bars and the eco-lodges. The cloud forest has over 600 direct and indirect employees.	Many of the traditional ways of life and working have been lost as people now work in the service industries to provide for the needs of the visitors. Local people might find that they have to hold a few jobs as they might be low paid or seasonal.

Table 8 *continued*

Socioeconomic positives	Socioeconomic negatives
Incomes in the area promote further developments in conservation and community projects in the area. These include comprehensive local environmental education and maintenance programmes. For example, one programme has existed for over 20 years where thousands of local students have been given classes on cloud forest ecology and environmental issues.	Increased numbers of visitors can also bring conflict and overcapacity. Tourists might come under the influence of alcohol or drugs and there might be an increase in the amount of antisocial behaviour or crime.
The cloud forest operates a lodge — La Casona — and they have achieved level 3 in the Costa Rican Sustainable Tourism index.	Greenwashing is a marketing scheme where a 'green' label is given to travel services that might not technically be classed as ecotourism. Ecotourism projects and resorts need to take care to make sure that they manage things carefully.

Summary

- Energy is transferred throughout an ecosystem through trophic levels. Nutrients mainly cycle between stores within the ecosystem.
- A large-scale threat in the tropical rainforest ecosystem is deforestation and the clearing, logging and cultivation for cash crops and animal grazing in the Amazon.
- The Amazon basin is one of the most diverse ecosystems on our planet but it has been under threat from the activities of humans.
- The Monteverde Cloud Forest in Costa Rica is an example of a small-scale case study that is trying to promote sustainable development within the tropical rainforest ecosystem. This will bring environmental and socioeconomic positives and negatives to the area.

■ Option C Dynamic coastal environments

Coastal processes and features

Coastal processes

Wave action

Waves are formed by the transfer of energy from the wind to the sea surface. As a breeze blows over the sea surface, it will cause ripples on the water. These little ripples act as 'sails' for the wind to blow against, transferring energy from the wind to the sea. They also cause eddies in the wind and these in turn stir up the water more, further encouraging the wave to grow.

Wave size

The size of the wave is influenced by various factors:
- **Wind strength**: the stronger the wind, the larger the wave.
- **Wind duration**: winds blowing over a longer period of time are better able to transfer energy to waves, making them larger.
- **Fetch** (the distance over which a wind blows to generate a wave): the longer the fetch, the greater the potential for waves to receive energy from the wind. For example, the west coast of Ireland has a fetch stretching far out into the Atlantic Ocean. The shores of Belfast Lough have a significantly shorter fetch.

Wave types

Winds produce two main types of waves:
- **Wind waves** are waves generated and influenced by the local winds blowing in the area in which the waves are occurring. They tend to have shorter wave lengths and faster wave periods.
- Once the wave-generating winds die down, the waves continue as **swell waves**. These waves lose very little energy as they travel and so can cover vast distances of thousands of miles. They tend to have longer wavelengths and longer wave periods.

Wave morphology

Waves can be described in the following ways (Figure 32):
- **wave crest** (highest point in the wave) and **wave trough** (lowest point)
- **wavelength**: the distance between two waves' crests
- **wave height**: the distance between the wave trough and crest.

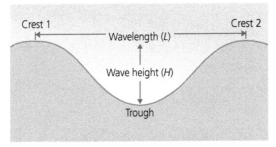

Figure 32 Wave morphology

Waves in deep water

In deep water, the motion of particles of water in the wave is circular. As the wave passes, the particles move forward and down, then backwards and up, tracing a circle and ending up at the same point at which they started (hence anything floating in water as a wave passes appears to bob up and down). This means that in deep water, the wave transports energy only, not matter. The diameter of these circles of rotation decreases with depth — the depth of the rotation extends to about half the distance of the wavelength, the circles of rotation becoming smaller in diameter as depth increases.

Waves entering shallow water

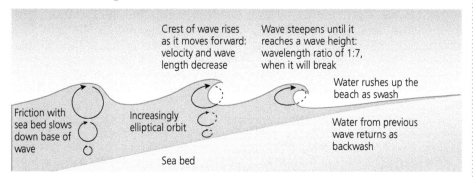

Figure 33 Waves entering shallow water

When a wave approaches the shore and enters shallow water, some significant changes occur (Figure 33).

- The bottom of the circular rotation of a wave crest touches the sea floor, experiences friction, and slows down. The rotation becomes more elliptical also as the motion of the wave particles becomes squeezed in the ever shallower water.
- The wave crest behind it is still travelling at the original speed, however, and so the wavelength decreases as it 'catches up' with the crest ahead. This concertina-ing of the waves forces the wave height to increase and the length to decrease.
- Eventually, the wave becomes so high and steep that the top of it spills forward and the wave **breaks**.
- As the wave spills forward up the beach (the **swash**), it now transports matter as well as energy — this includes beach sediment, and any surfer who has managed to catch the wave! Eventually, the swash reaches the furthest point in its journey up the beach and moves back down the beach as **backwash**.

Types of breaking waves

Waves can be divided into two main types: constructive or destructive (Figure 34 and Table 9).

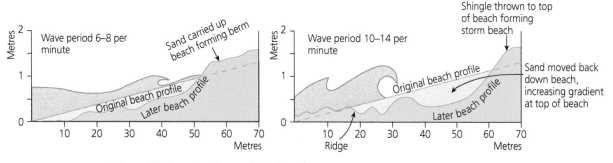

Figure 34 Constructive and destructive waves

Exam tip

Be clear on how a wave changes as it moves from deep to shallow water.

Knowledge check 14

What are the key changes that occur to a wave as it enters shallow water?

Table 9 Constructive and destructive waves

	Constructive	Destructive
Conditions likely to produce them	Gentle gradient beaches during less stormy conditions (i.e. the summer)	Steeper beaches during more stormy conditions (i.e. winter)
Wave characteristics	Long wavelength and low wave height	Short wavelength and higher wave height
How the wave breaks	Constructive waves produce spilling breakers with a strong swash and weak backwash	Destructive waves produce plunging breakers with a weak swash and strong backwash
Impacts on the beach	Material is moved from the bottom to the top of the beach by the strong swash, increasing beach gradient. This produces **berms** at the top of the beach, and also a series of **ridges and runnels** (depressions) running laterally along the beach.	Material is moved from the top to the bottom of the beach, decreasing beach gradient. In fact, the material may be removed entirely from the beach on a temporary basis during the winter, stored on an offshore bar, and brought back in the following spring and summer by constructive waves. As the higher energy plunging breakers break, they may also throw some shingle above the high water mark, forming a **storm beach** at the top of the beach.

It should be apparent from Table 9 that constructive waves will lead to the conditions more suited to destructive waves and vice versa. In fact, it is often possible to see an annual cycle on beaches: lower, steeper beaches during the winter due to the destructive waves, followed by higher, less steep beaches in the summer, resulting from constructive waves.

Knowledge check 15

Distinguish between constructive and destructive waves.

Wave refraction

As waves approach a coastline, they tend to refract (bend) and become increasingly conformed to the shape of the coastline. This is due to frictional drag with the sea bed as the water gets shallower (Figure 35).

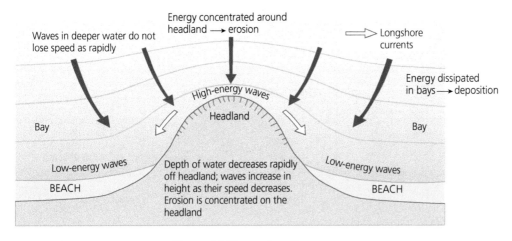

Figure 35 Wave refraction

This has an important consequence for energy distribution along coastlines. As the waves refract around a headland, their energy tends to become more concentrated and hence they experience more erosion. On the other hand, as they spread out into a bay, their energy is more widely dissipated, hence bays become places dominated by deposition.

Wave erosion

Waves erode coasts through the following processes:

- **Hydraulic pressure**: as a wave breaks against a coast, it may trap air in crevices in rocks. This increases the air pressure, putting stress on the rock. Over time, this weakens the rock and may cause it to break and erode.
- **Abrasion/corrasion**: as waves break, they are able to transport the sand and shingle along the shore line. As the sand is moved across the surface of rock, it can abrade the rock and erode it. Larger waves may fling or knock pebbles against the rock, again causing it to erode.
- **Wave pounding**: the sheer mass and movement of large waves as they break can cause considerable amounts of energy to be transferred onto the coasts, producing significant erosion.
- **Corrosion/solution**: carbonic acids in sea water can dissolve limestone and chalk (such as the cliffs found at Whiterocks near Portrush). In addition, the salt in the water can deposit salt crystals on the rocks. As these grow and expand, they can cause erosion of the rock.
- **Attrition**: the sediment previously eroded can hit off other sediment, causing it to become smaller and more rounded.

Wave transport

We have already discovered that waves can transport matter as soon as they break. There are various forms of transportation:

- **Solution**: the dissolved material is transported by the moving water.
- **Suspension**: the smaller material is carried by the turbulence of the breaking wave in both swash and backwash.
- **Saltation** (bouncing) and **traction** (rolling): the largest material, including shingle and pebbles, is transported by being bounced (lifted up before being quickly deposited again) or rolled by the breaking wave.

In addition, sediment at coasts can be transported by the wind. Strong winds can move sand across the surface of the beach by suspension, saltation and traction.

Swash-aligned and drift-aligned beaches

Swash-aligned beaches

These occur where waves break parallel to the shore line and the swash and backwash move vertically up and down the beach. This occurs as wave refraction bends the waves to conform to the shape of the beach. Here, the predominant movement of sediment is either onshore via constructive waves or offshore by destructive waves.

Exam tip

Make the link between wave refraction around headlands and bays with the coastal features found in Figure 38 on p. 55.

Drift-aligned beaches

These form where waves break at an angle to the beach (this tends to happen where offshore and nearshore gradient is steep and waves do not have time to refract around to break parallel with the coast). Here, another vital coastal transportation process occurs: **longshore drift** (Figure 36). As the wave breaks at an angle, the swash moves up the beach at the same angle. Any sand or shingle carried by the breaking wave thus moves up the beach at the same angle. However, the backwash follows the most efficient way back down the beach, and so moves back down straight. Again, this carries the sediment back with it. The end result is that the sediment moves along the beach with each breaking wave, following a zig-zag route.

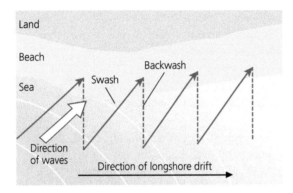

Figure 36 Longshore drift

Formation of landforms at high-energy and low-energy coasts

Landforms at high-energy coasts

We have seen how wave refraction concentrates wave energy around headlands (promontories of harder, more resistant steeper-sided coastline that protrude out seawards from the bays in between them). This produces a characteristic series of features.

Exam tip

Make sure you can refer to places to give examples for each of the features.

Cliffs

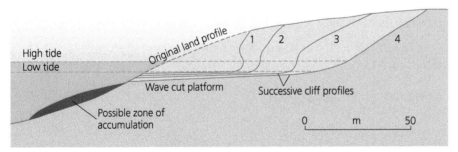

Figure 37 The formation of cliffs

Coastal cliffs are slopes found at the shoreline where wave erosion plays an important role in their development (Figure 37). As waves collide against a slope, they erode the base between the high and low water mark. This forms a **notch** on the shore line and undercuts the material above it. As this material is unsupported, it will eventually collapse, leaving behind a straighter, steeper section of shore, and beginning the process of forming a **cliff**. Another notch is then carved out; the unsupported material above again will collapse, and the cliff grows in size. Over time, as this process repeats, the cliff will grow bigger and bigger as the waves erode further inland. If the cliff is formed from rock, this process leaves behind a flat platform of rock, revealed during low tide, known as a **wave cut platform**. Many sections of the North Antrim coast have cliffs made from the basalt laid down during the formation of the Antrim Plateau.

The form the cliff takes is influenced by various factors. For example, if the bedding planes of the rock dip towards the sea, the cliff angle will be less steep than places where the bedding planes are more horizontal or dip away from the coast.

Headlands

Headlands are protrusions of more resistant rock out from the shoreline. They can be formed where you have rocks of resistance beside one another running perpendicular to the coast. For example, Ramore Head at Portrush was formed where a hard, resistant headland of dolerite protrudes out into the sea.

They can also occur where rocks of different strengths lie parallel to the shore. In this case, a fault in the more resistant rock at the coast may allow erosion through to the weaker rock behind. This may erode out to form a bay, leaving headlands of the harder rock. This produced Lulworth Cove, Dorset.

Caves, arches, stacks and stumps

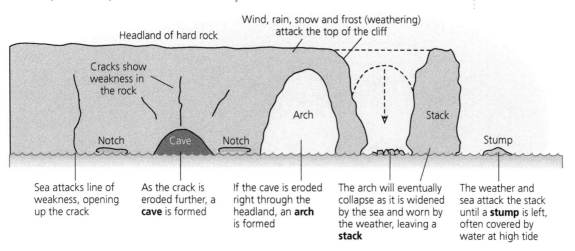

Figure 38 Headland features

As the waves break around a headland, they will preferentially erode any places of weakness. These include fractures or joints in the rock. These will be expanded out by continuous erosion to produce a **cave**. Where that line of weakness carries right through the headland to the other side, a cave will be formed here also. Over time,

> **Exam tip**
>
> Practise drawing well-annotated diagrams to illustrate the formation of each of the features. The annotation should focus on explaining the processes rather than simply labelling the features.

as the caves get bigger, they may eventually meet in the middle of the headland. The material in the roof of the cave is unsupported and is prone to further erosion and collapse. So, eventually, the cave may be widened out to form an **arch** right the way through the headland. The unsupported roof of this arch will eventually collapse also, leaving behind an isolated pillar of rock known as a **stack**. The base of the stack is eroded by breaking waves, carving out a notch around it. This undermines the stack, causing it to collapse. Eventually, the only thing left is a submerged **stump** of rock. There are examples of caves, arches and stacks found in the chalk cliffs of Whiterocks at Portrush.

Landforms at low-energy coasts

Just as wave energy is concentrated at headlands, so it is dissipated in low-energy environments such as bays. In these locations, depositional features dominate.

Beaches

Beaches are depositional landforms found in low energy intertidal zones at coasts. The sediment to form a beach can come from various sources:

- erosion of cliffs
- longshore drift
- deposition of material from nearby rivers
- offshore sources (by constructive waves)

Where the rates of deposition of this material are greater than the rates of transportation away, a beach will build up (Figure 39).

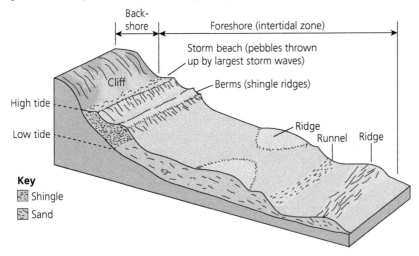

Figure 39 Beach features

Sandy beaches are often much more gentle than shingle beaches and tend to be well sorted (i.e. made up of particles of very similar size). This is because of the action of swash and backwash. The spilling swash of constructive waves moves the sediment up the beach; however, the weaker backwash is only able to remove the smaller sediment, carrying it back down to the beach. Consequently, the coarsest particles are found towards the top of the foreshore, while the smallest particles are found in the offshore zone. This also explains why, on beaches that have both sand and shingle, the shingle is typically found at the backshore and the top of the foreshore, while the sand is found closer to the nearshore and offshore.

Sandy beaches can be shaped into a series of **ridges** and **runnels** (depressions), large undulations across the surface of the beach up to 100 m from crest to crest and up to 1 m in height. These run perpendicular to the shore.

There are many fine examples of beaches around the coasts of Northern Ireland, including Murlough in Co. Down and Portstewart Strand.

Dunes

Dunes may form where the following conditions are found (Figure 40):
- large supply of sand and sediment (e.g. from a beach)
- sufficiently energetic winds that are onshore for periods of time
- large area of flatter land behind the shore

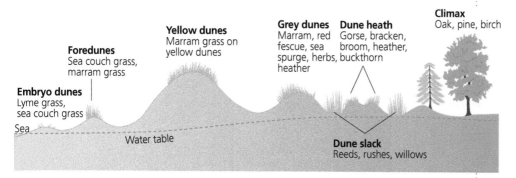

Figure 40 Sand dune features

Where these conditions exist, dunes are formed as follows:
- Sand found above the high tide level is transported by onshore winds towards the back of the beach. Where there are obstacles such as pieces of driftwood or seaweed, the sand can build up in height, forming the **embryo dunes**.
- This increased height raises the beach up slightly away from the influence of the tides and allows plant succession to begin. Salt tolerant plants such as sea couch colonise the embryo dunes. Their roots help stabilise the sand, and further deposition occurs around the long, thin leaves of the grass, further raising the height of the dunes, and forming the **foredunes**. Foredunes can build quickly as plants like marram grass begin to grow, forming a height of several metres in 5–10 years.
- Improving soil conditions, height and stability of the dunes encourage plant succession to occur, eventually forming first the **yellow dunes,** then **grey dunes** (fixed dunes with a layer of organic matter, and with red fescue replacing marram grass) followed by **dune heath** (the area behind the grey dune where the supply of beach sand is reduced, resulting in smaller dune features).
- The parallel ridges in the dunes are interspersed by lower lying **dune slacks**.
- Behind the dunes, **climatic climax** may be reached as deciduous trees grow.

The outcome of these processes can be seen in the sand dunes at Portstewart Strand.

Exam tip

See Student Guide 1 in this series, pp. 42–43, for more details on sand dunes.

Given their highly dynamic nature, dunes are a vital part of the natural coastal system. When beaches lose sediment via destructive waves in the winter, wind and wave erosion from the front of the dunes can transfer material onto the beach, supplementing it during times when it would otherwise be lower. The embryo dunes can be built up again during the following summer as sand carried back into the beaches by constructive waves can be blown towards the back of the beach by the wind. So when dunes are removed by humans, for example to build a sea wall, this can have devastating consequences for beaches.

Spits

Spits are linear ridges of deposited sediment that extend out from the coastline across, for example, an estuary mouth (Figure 41).

There are two characteristics that are required for their formation:

■ an abrupt change in the orientation of the shoreline, for example at an estuary mouth
■ a drift-aligned beach with longshore drift occurring

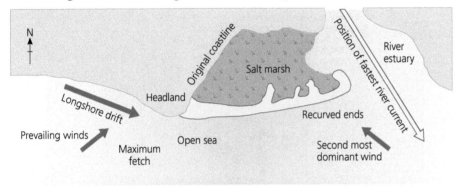

Figure 41 The formation of a spit

If these conditions are found, the following processes form the spit:

■ The prevailing wind causes longshore drift to carry sediment along the coastline. When the sediment reaches the estuary mouth, however, rather than following the new direction of the coast, it continues to travel in the original direction. This causes sediment to be carried out and deposited across the mouth of the estuary. Storm waves throw material above the high water mark, building the spit up above sea level.

■ A hook may form from wave refraction around the end of the spit. This causes some material to be carried around its end and deposited just in behind it. This process can be aided by storm waves coming from an alternative direction. These waves carry material around the end of the spit, building up the hook. In fact, a spit may have a series of hooks marking the former end positions of the spit as it was developing.

■ The spit may stabilise over time if dunes are formed towards its back.

■ Behind the spit, a low energy environment is created. Here, silt and mud are deposited forming a **salt marsh**. These are intertidal mud flats that can build up over time so that they may only be covered by the sea for an hour per high tide. In fact, the highest points in the salt marsh may only be covered by the highest of spring tides. Salt marshes have often been drained and reclaimed by humans for agriculture.

There is a spit at Larne which formed following the end of the last glaciation. A plentiful supply of sediment was carried to a change in orientation of the coast, building outwards from the shore. This has now largely been built upon following the development of Larne Harbour.

Bars

If a spit forms completely across a bay, joining the two headlands, the feature formed is called a **bar**. This may occur if there is not significant outflow of water from a major river into the bay. Given the lack of water flow, the lower energy environment allows deposition to occur right across the bay, forming the bar.

Tombolos

These are specific beaches which form out from the shore and join an island to the shoreline (Figure 42). They are formed by the action of wave refraction around the island. As the waves approach the island, they refract around it. This means that the wave crests approach each other head-on around the back of the island. As these waves break on the beach from opposing directions, they cause longshore drift to move from these two directions and hence towards each other. Where these drifts meet behind the island, the sediment starts to build out seawards. In effect, the sand is swept together by the approaching waves. Over time, it builds upwards and outwards, eventually joining the island to the shoreline forming the tombolo.

There is a small example of a tombolo at Ballintoy Harbour. A sudden change in the orientation of the shoreline near a headland combined with offshore stacks caused wave refraction to form a small tombolo, extending the beach out from the shore to join on to the stack.

> **Knowledge check 16**
>
> Distinguish between a spit and a bar.

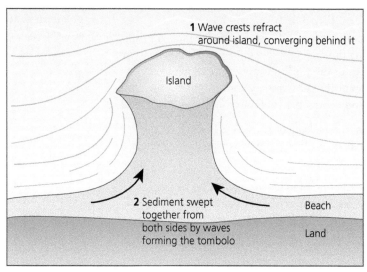

Figure 42 The formation of a tombolo

Summary

- As waves move from deep water into shallow water, they begin to break and then they transport matter as well as energy.
- Constructive waves occur on gentler beaches, have spilling breakers and tend to move the sediment up towards the top of the beach. Conversely, destructive waves have plunging breakers and tend to move material down the beach towards the offshore zone.
- As waves approach the shoreline, they tend to refract around the shape of the shoreline. This causes them to wrap around headlands, concentrating erosion there.
- Longshore drift is an important coastal process. It moves sand and shingle along the coastline.
- Erosion at the coast produces various landforms, including cliffs, headlands, caves, arches, stacks and stumps.
- Coastal depositional features include beaches, dunes, spits, bars and tombolos.

Regional coastlines

Processes and features associated with coastlines of submergence and emergence

Processes of submergence and emergence

We are used to the fact that the sea level changes on a day-to-day basis due to tides. However, sea levels can rise and fall on much longer timescales too. These changes can be split into two categories.

Eustatic (global) changes in sea level

Eustatic changes are due to changes in the volume of water in the oceans, for example during a period of glaciation and during interglacials. During a period of glaciation there is a fall in sea levels because more of the world's water is stored in ice sheets, and because of the thermal contraction of the seas and oceans as temperatures fall (the cool water becomes more dense and contracts in volume).

Conversely, during interglacial periods, the melting of ice sheets (especially the continent-based sheets) and the thermal expansion of the oceans cause sea levels to rise.

Around 18,000 years ago, when the last period of extensive glaciation was at its maximum, sea levels were around 120 m lower than present sea levels. Between 18,000 years ago and around 7,000 years ago, sea levels rose eustatically.

During the past century, due to climate change, thermal expansion and melting of continental glaciers has caused global sea levels to rise eustatically by 18 cm.

Isostatic (local/regional) changes in sea level

Glacial isostatic changes

Isostatic changes can also be related to glacial and interglacial periods. During times of glaciation, large ice sheets can occupy some land masses. For example, during the Pleistocene, much of the British Isles was covered by ice sheets, some up to 3 km

> **Exam tip**
>
> Make sure you fully understand and can explain the differences between eustatic and isostatic changes.

thick. The weight of this ice compressed the lithosphere downwards on top of the asthenosphere, displacing the mantle rock laterally, and lowering the level of the land. This effectively led to local sea level rise in the British Isles. As the last period of glaciation began to end around 11,700 years ago, the ice melted and the British Isles began to rise up again isostatically. Isostatic rise was initially relatively fast due to the removal of the mass of the ice. Following this, the rising continued but at a much slower rate, as rock in the asthenosphere slowly flowed back in below the lithosphere.

Tectonic isostatic changes

Isostatic changes may also be associated with tectonic activity. For example, during the 2011 earthquake that produced the devastating tsunami in Japan, some of the coastal areas in Japan sank. Tragically, this allowed the tsunami to overtop some of the sea walls. This occurred as the sea floor shunted suddenly upwards near the Tohoku fault. As this part of the sea floor rose, it caused the coastal regions of Japan to sink. For example, the coastal plain of the Sendai region near the epicentre dropped — before the quake, around 300 hectares were below sea level and after the quake, this had risen to nearly 16,200 hectares.

Isostatic change can also occur due to longer-term tectonic processes. The process of **orogeny** (mountain building) at collision zones can cause localised uplifting of rock, for example in the Himalayas and the Alps.

Landforms of submergence and emergence

As we move out of a period of glaciation, there will be both global eustatic and regional isostatic changes occurring. Eustatically, global sea levels will be rising due to melting of continental ice sheets and thermal expansions of the oceans, producing a positive change in base level. At the same time, local isostatic rebound will occur as the ice sheets melt and the lithosphere begins to rise, producing a negative change in base level.

As the British Isles came out of the last period of extensive glaciation, at first sea levels rose due to eustatic change. This change in sea level finished around 4,000 years ago. Isostatic rebound occurs more slowly, however, and the northern half of the British Isles is still rising today due to this process. The balance between the rates of these two changes (positive changes in base level resulting from eustatic fall in sea level and negative change in base level resulting from isostatic rebound) can produce different landforms.

Submergent landforms

Where rates of isostatic rebound are less than the rates of eustatic rise, there will be a net rise in sea level as the level of the land rises more slowly than sea level is rising. This produces the following features:

- **Rias** are formed when unglaciated river valleys become flooded due to sea level rise. The floodplain of the drainage basin becomes flooded and they often produce a tree-like pattern corresponding to the various tributaries that fed into the main river. Lough Swilly in Donegal is an example of a ria.
- **Fjords** form when the sea level rise floods a glaciated river valley. Having been glaciated, the river valley is classically U-shaped, and so fjords are inlets characterised by very steep-sided valleys and form long, deep and narrow inlets. At the upper end of a fjord, the river flowing into it may form a small delta. Milford Sound in New Zealand is an example of a fjord.

Knowledge check 17

Distinguish between eustatic and isostatic changes in sea level.

Exam tip

The concept of **base level** is useful here. This is the lowest level to which erosion by running river water can occur and so corresponds with sea level. When sea levels fall, there can be said to be a negative change in base level and vice versa.

Emergent landforms

Where rates of isostatic rebound are greater than the rates of eustatic rise, there will be a net fall in sea level as the level of the land rises faster than sea level is rising (Figure 43). This produces the following features:

■ **Raised beaches**: as the name suggests, these are former beaches (and wave cut platforms) that have been raised up above the current sea level. They can be identified by the presence of a former sea cliff with a former wave cut platform in front of them. The road out from the east of Portrush towards Whiterocks runs along a raised beach. Landward, there is a cliff marking the former location of the shore.

■ **Relict landforms**: in addition these cliffs may contain a number of coastal landforms that have also been raised above current sea levels, including wave cut notches, caves, arches and stacks. The caves at Ballintoy Harbour are examples of relict coastal landforms, now raised above sea level.

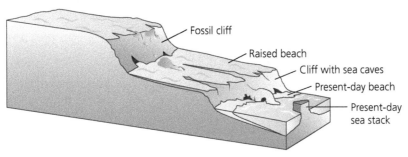

Figure 43 Features of an emergent coastline

Knowledge check 18

Explain how eustatic and isostatic changes can interact to produce emergent landforms.

Threat of rising sea levels due to climate change on the human and physical environment

During the twentieth century, global sea levels rose by an average of $1.8 \, \text{mm} \, \text{year}^{-1}$. However, closer analysis of the rising trends show that the rate of rise is increasing: $0.8 \, \text{mm} \, \text{year}^{-1}$ between 1870 and 1924, $1.9 \, \text{mm} \, \text{year}^{-1}$ between 1925 and 1992, and $3.1 \, \text{mm} \, \text{year}^{-1}$ from 1993 to 2012.

This rise has been due to climate change as global temperatures have risen. Rising temperatures cause sea levels to rise in two ways:

■ **Thermal expansion of the oceans**: as the water warms, it expands and causes sea levels to rise.

■ **Melting of land-based glaciers and ice sheets**, adding more water into the oceans.

Exam tip

Find out more details on climate change and sea level rise on pp. 83–86. Even if you are not studying it as one of your optional units, it will deepen and broaden your geographical understanding.

Exam tip

As always, be able to refer to places to illustrate the features. These may be required to enable you to achieve top marks in the exam.

Case study

Sea level rise in Kiribati

Physical environment

More frequent and extreme coastal flooding

On average, it is estimated that Kiribati has experienced an average sea level rise of around 3.7 mm since 1992 (this is hard to estimate given that El Niño years can produce temporary sea level rises). However, it is the extreme events that accompany climate change and sea level rise that currently are impacting Kiribati. For example, in 1997 Kiritimati island was devastated by an El Niño event that brought heavy rainfall and flooding, resulting in a half-metre rise in sea level. Roughly 40% of the island's coral died and their 14 million birds left the island. The 'king tide' event that inundated Kiribati in 2005 produced a storm surge of 2.87 m, which destroyed some villages, swept crops out to sea and contaminated water supplies.

In March 2015 Kiribati experienced flooding and destruction of sea walls and coastal infrastructure as the result of Cyclone Pam, a category 5 cyclone that devastated Vanuatu. Kiribati remains exposed to the risk that cyclones can strip the low-lying islands of their vegetation and soil.

Contamination of fresh groundwater

The islands consist mostly of coral limestones. These rocks are porous and allow sea water to infiltrate easily. This means that the salt water table rises and falls with each tide — but also is rising incrementally as sea levels rise. The main consequence of this is the contamination of many of the island's wells. The overall volume of groundwater in South Tarawa is dependent on the size of land area of the island. As this shrinks, so the volume of fresh groundwater is reduced.

The coral limestone is porous and allows sea water to flow through it. The saltwater table oscillates on a daily basis with the tides, and in the long term with the mean sea level. As sea levels have risen, many wells have become contaminated with salt water and can no longer be used.

Loss of land

Some of the smaller islands of Kiribati have already been lost to sea level rise. Two small uninhabited Kiribati islets, Tebua Tarawa and Abanuea, disappeared underwater in 1999.

The islands of Kiribati are coral atolls. Coral can respond to sea level rise — if it is gradual enough, the coral pulyp can raise the atolls up as sea level rises. In fact, there is some evidence of expansion on land in some of the islands of Kiribati. A study by scientists from New Zealand found that three of the main urbanised islands — Betio, Bairiki and Nanikai — increased by 36 hectares, 6 hectares and 0.8 hectares respectively due to the growth of new coral. However, it stressed that the study examined area only and not vertical growth of the islands, and it concludes that the vulnerability of islands remains high and that even these new corals would be vulnerable to sea level rise and inundation. Also, if sea level rises at a rate faster than coral growth (and if the coral is damaged by ocean acidification, another consequence of global increases in carbon dioxide), then the islands continue to be at risk.

Exam tip

If writing about this in the exam, make sure your answer is well balanced between the physical and human impacts. Imbalance in your answers is one of the things that can take you out of Level 3.

Human environment

Economic costs

Unless significant adaptation measures are introduced, Kiribati's capital of Tarawa, home to half the country's population, will be 25–54% inundated in the south and 55–80% in the north by mid-century, and the country's one airstrip in the south of the island will be eroded. The causeway roads that link the various parts of Tarawa together are being eroded by rising seas. If damaged, this could disrupt socioeconomic links, making it difficult for inhabitants to access services including the island's hospital. For a country with an annual GDP of only $600,000, the cost of the adaptation measures needed for all inhabited islands is estimated to be around $2 billion.

Food security problems

Coconut is the main cash crop for about 55% of the population. But it and other crops such as breadfruit and giant taro are susceptible to saltwater intrusion due to storm surges and overwash. On the island of Abaiang, the freshwater milkfish that provided protein to the local people are now totally gone. The taro plant, which is grown in groundwater pits 200 m from the coast, is threatened due to saltwater contamination of the groundwater supplies. In 2013, the government of Kiribati purchased 20 km² of land on Vanua Levu, one of the Fiji islands, around 2,000 km away. It bought the land from the Church of England for around $9 million. The land is currently being used for farming and fishing to try to guarantee the nation's future food security.

Relocation of villages within Kiribati

The populations of many of the villages closest to the sea in the islands have had to relocate. For instance, the village of Tebunginako on the island of Abaiang has already been badly affected by rising seas. Over the past 40 years, sea level rise and storm surges have caused major issues for the inhabitants of Tebunginako. Groundwater began to become contaminated in the 1970s. The loss of land became so great that the village had to be relocated on the island. Now, the remains of thatched huts and the village hall sit around 30 m offshore. Around 2005, the village was relocated about 50 m from the then position of the shore. However, now at high tide, some of the houses, the Catholic church and village hall are surrounded by a moat of saltwater as the sea flows in and floods what was formerly a freshwater pond.

On the main island of Tarawa, 40 homes in the village of Te Bikenikoora have had to be relocated up to 2013 and the meeting house was only saved from a king tide in 2013 by the actions of all the villagers.

Summary

- Sea levels can change due to eustatic and isostatic causes. Eustatic changes are global changes in sea level associated with glacial and interglacial periods. Isostatic changes are local or regional changes in sea level resulting from glacial and tectonic causes.
- These sea level changes produce associated landforms. Emergent landforms include raised beaches and relict landforms. Submergent features include rias and fjords.
- Rising sea levels associated with climate change threaten the physical environment and the human environment, as can be seen in the case study of Kiribati.

Coastal management and sustainability

Role of EIAs, CBA and SMPs

Coastal management must strive for a difficult balance:

- protecting social and economic assets at a particular stretch of coast, at a reasonable cost, and using strategies that minimise environmental impact in the area in which they are used
- all of this is done in a broader context: the actions taken at one location should have minimal negative impacts in other locations in a connected coastal system

■ and the actions should be sustainable in the long term of decades and into the next century

To provide a framework to guide such complex decision making, **Shoreline Management Plans** (SMPs) are used in the UK. They guide sustainable coastal management into the twenty-second century. SMPs seek to:

■ identify the social, economic and environmental risks of flooding and coastal erosion in a sediment cell

■ set out the preferred policy approach to address these risks

■ and outline the consequences of the approach recommended

In doing so, SMPs consider three key factors (Table 10).

Exam tip

Read more about the concept of climate change adaptation on pp. 87–88.

Table 10 Factors considered in SMPs

Factor	Explanation
Cost–benefit analysis (CBA)	This approach weighs up: Benefits: what is the value of the land being protected in terms of human infrastructure (including settlements, industry, power stations) and economic activities (such as tourism)? Costs: what is the cost of building and maintaining coastal defences? If the costs outweigh the benefits, coastal protection is less likely to be recommended.
Environmental Impact Assessment (EIA)	What are the impacts of action (or inaction) on the local ecology and habitat diversity? To what extent does the local geology and wave energy environment make the area susceptible to erosion? What is the likely impact of climate change on sea levels in this location?
Technical viability	If it is decided that a section of coastline merits active management intervention, to what extent can the proposed management strategies actually produce the desired management outcome? And what about its potential impacts on other sub-cells of the sediment cell? In short — will it work? For example, protecting a short section of coast while allowing sections either side of it to erode could have the longer-term impact of creating a prominent headland at the protected site. This will not only reduce the chances of allowing a beach to continue to exist at the site, but it could interrupt the natural flows of sediment along the entire coastline, trapping it in the newly formed 'bays' either side of the protected site. This approach would not be very sustainable and would be less likely to be approved.

There are four broad policy approaches SMPs may choose from (Table 11).

Exam tip

The specification explicitly mentions CBA and EIA, so make sure you can discuss these in detail.

Table 11 SMP policy approaches

Approach	Explanation
Hold the line	Maintaining the current coastal position by repairing or upgrading existing defences.
Advance the line	Placing new defences seaward of the original defences (rare).
Managed retreat	Allowing coastal erosion processes to occur at a location while implementing management strategies to manage the retreat of human and economic assets.
No active intervention	No action taken other than monitoring developments.

Exam tip

See the case study on north Norfolk on pp. 68–70 for examples of these different policy approaches.

Evaluation of hard and soft engineering strategies on the human and physical environment

Once a decision has been made in the context of the SMP regarding which of the four policy approaches is to be followed, various hard and soft engineering strategies may be used to implement the policy.

Hard engineering strategies

Hard engineering strategies try to control natural processes by putting intervention strategies into place that modify or work counter to natural processes. As such, they are generally less sustainable in the long term. However, where high value areas are under threat, and where their use is technically viable, hard engineering strategies can still be employed as part of SMPs (Table 12).

Table 12 Hard engineering coastal management strategies

Strategy	Impacts	Sustainability
Sea walls Recurved sea wall — Concrete — Beach material — Steel pile	Absorb and reflect wave energy so stop coastal retreat	■ As they stop erosion of the shore, they reduce the input of sand into the coastal system, leading to negative impacts elsewhere along the coast. ■ Expensive, both to build and to maintain. ■ Their sloped design reflects the waves, encouraging backwash. This can lead to beach erosion and loss of the beach entirely in some cases.
Revetments: open sloped structures designed to absorb wave energy Wooden revetment — Open structure of planks to absorb wave energy but allow water and sediment to build up beyond	Also absorb wave energy, reducing the erosion of the coast or cliffs behind	■ They are cheaper than sea walls, but are often less robust and so will need more regular maintenance. ■ They allow the water to break through them so they are not as prone to encouraging backwash as sea walls. ■ They are visually unpleasant and long lines of revetments along a shore can restrict access to the beach for people.
Rip-rap: large boulders placed along the shore Rip-rap — Large boulders dumped on beach	Designed to absorb the wave energy of breaking waves	■ They are more robust than revetments and allow some percolation of water, so they do not encourage the same backwash as sea walls. ■ They are visually imposing and they can create a physical barrier making access to the beach more difficult for people.

→

Table 12 *continued*

Strategy	Impacts	Sustainability
Gabions: metal cages filled with boulders Gabion — Steel wire-mesh cage filled with small rocks	Also absorb energy of the breaking waves	■ Relatively cheap. ■ Allow some percolation of water, so they do not encourage the same backwash as sea walls. ■ The cages are vulnerable to splitting as they absorb the energy, requiring ongoing maintenance.
Groynes: barriers (wooden, concrete or boulders) placed perpendicular to the shore Groyne — Wooden or steel piling — Concrete wall — Beach material	Interrupt longshore drift, trapping beach material in the location where the groynes are	■ By increasing the beach size, the beach acts as a buffer between the shore and the breaking waves. The beach can respond dynamically to changing wave environments. ■ Locations down the shore from the groynes are deprived of beach material, increasing the erosion risk there.

Soft engineering strategies

Soft engineering strategies attempt to work with natural processes, basing interventions on strategies that harness natural processes in the defence of the coast (Table 13). As such, they tend to be much more sustainable in the longer term; however, these strategies can involve negative impacts for areas that are allowed to erode as part of a managed retreat policy in order to provide sediment supplies for elsewhere in the sediment cell.

Table 13 Soft engineering coastal management strategies

Strategy	Impacts	Sustainability
Beach nourishment	By building up the beach, it then acts as a buffer between the shore and the breaking waves. The beach can respond dynamically to changing wave environments and can protect the shoreline behind it.	■ This is a softer engineering approach as it uses natural processes to manage the erosion threat. ■ However, it does not tackle the cause of loss of beach material, and so ongoing and expensive maintenance will be required. ■ The costs of beach nourishment are high. And the challenges of finding a suitable source of sediment to add to the beach are great.
Dune regeneration	A functioning, natural dune can act as a source of sand for a beach that is being eroded by destructive waves in the winter.	■ This is a very sustainable approach as it allows natural transfers from the dune to the beach and vice versa, helping maintain the beach as a defence against sea erosion. ■ It also increases habitat diversity.
Managed retreat	The overall impact in a sediment cell is positive, in that the area being eroded can naturally provide a supply of material to build up beaches elsewhere. However, the short-term impacts of loss of land and infrastructure can be considerable on those directly affected.	■ Overall, this is a very sustainable strategy as it allows a sediment cell to operate in a natural way, inputting sediment into the cell which will help replenish beaches elsewhere. ■ There are limited ongoing maintenance costs (no hard engineering is involved). ■ It allows a coastal system to find a natural dynamic balance.

The SMP outlines the following impacts that should be considered in making a decision on policy approaches:

- **Physical environment:** there is a legal requirement to consider the environmental impacts of interventions in Special Protection Areas (SPA) or Special Areas of Conservation (SAC) and government policy requires local councils to avoid environmental damage and to look for opportunities for environmental enhancement.

- **Human environment**
  - **Land use**: in the past, there has been unhindered development of settlements and other human activities along the coastline. However, this has resulted in the need for hard engineering protection, which is not economically sustainable in the longer term.
  - **Tourism/recreation**: although tourism to coastal areas has declined in recent decades in the UK, tourism is still an important part of the economies of many coastal locations. So, the SMP needs to consider the impacts on the tourism industry when reaching policy decisions.
  - **Heritage**: there are many places of heritage value along the coast. The SMP should, where possible, seek to protect these. But it is recognised that this is not always possible or sustainable. And so each location should be considered on a place-by-place basis — which may mean decisions are made that result in the loss of heritage sites.
  - **Communities**: indices of deprivation show that many of the UK's coastal areas have issues such as unemployment, health problems and the impacts of seasonal visitor business. These problems can be worsened when the communities face threats from coastal erosion or flooding.

Knowledge check 19

Explain why soft engineering strategies are generally preferred over hard engineering strategies.

Case study

North Norfolk

Context
- Geology: the cliffs of the north Norfolk coastline are mostly made up of deposited glacial till, which is very susceptible to erosion.
- This part of Britain is experiencing isostatic rises in sea levels, as it sinks following the end of the last glaciation, combined with the broader eustatic rises in sea level associated with climate change. This increases its vulnerability to coastal erosion.
- The north Norfolk SMP identifies evidence that this shoreline has been retreating for centuries and will continue to do so into the future. It states that 'Human intervention will not halt this natural process (of erosion); coastal defence works carried out over the last century have not prevented natural change from occurring, they have simply delayed its full implications from being felt.' It goes on to say that the 'approach to resist erosion and shoreline retreat ... is only sustainable for short periods of time. The decision to be made now is how we are going to manage this natural change in the future.' (SMP p. 9)

In this context, we will examine a range of hard and soft engineering strategies used at various locations along this SMP shoreline and evaluate their impacts and sustainability.

→

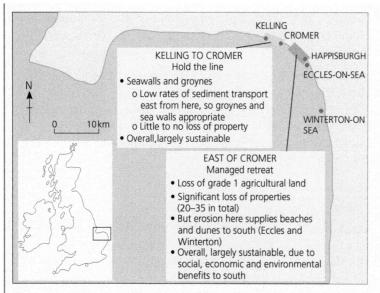

Figure 44 Coastal management in north Norfolk

Kelling to Cromer

Hold the line via sea walls and groynes

Cromer has had a history of coastal erosion. Large portions of the cliffs were undermined and overwhelmed in 1825 and 1832; and a jetty, which gave the only harbour accommodation, was washed entirely away in 1845. A sea wall was first built in 1847 at a cost of nearly £10,000. This sea wall is now backed either by regraded vegetated cliffs or a large additional wall, extending approximately 25 m high. The sea wall acts to absorb and reflect the waves' energy, thus protecting the position of the coastline and the important economic assets that are located there. A series of five groynes work to interrupt longshore drift, thus helping maintain the beach in front of the sea wall. The SMP policy here is hold the line.

Table 14 Impacts on physical and human environments

Physical environment	Human environment
As hard engineering strategies, normally sea walls and groynes would be expected to have a largely negative impact on other coastal areas, and they tend to reduce and impede sediment supply and transport. However, in the case of Cromer, the North Norfolk SMP notes that this section of coastline 'is characterised by low rates of sediment transport and relative stability when compared to much of the rest of the SMP coastline. Furthermore, the eroding cliff provides little contribution to beaches beyond.' Given this, the negative impacts to the environment have been minimal. By 2050, landscape character of seafront may change due to greater defence works. By 2105, beach may be lost.	By 2025, there will be no loss of property or land behind the existing defences, no loss of heritage sites and no loss of community or recreational facilities landward of defences. By 2055, it is still expected that there will be no loss of property, land or heritage behind the existing defences. However, properties along the promenade may become more exposed and subject to overtopping and storm damage. The structural integrity of Cromer Pier will possibly be threatened. The Lifeboat Station may need to be relocated.

Sustainability

Overall, the use of sea walls and groynes could be considered largely sustainable in this particular case. This is mainly because this section of coast has relatively low sediment transport rates and so stopping erosion at Cromer will have limited negative impacts elsewhere in the sediment cell. Additionally, the →

town is a main urban centre for the region, providing services that support a number of surrounding communities. It is envisaged that the groynes will help maintain the beach for around 20–50 years, and this will help protect the sea wall from further damage.

However, there are ongoing economic costs in the longer term in terms of maintenance costs for the sea walls and groynes. In fact, after around 50 years, the SMP predicts that the beach will no longer be in existence at Cromer, due to factors such as sea level rise, increased storms, and the fact that the sections of coast to the west and east of Cromer will continue to erode, leaving it as a promontory exposed to further erosion. Beyond 100 years, the SMP considers that the erosion pressures on the town will be such that it may be hard to justify the costs of maintaining the hard defences.

East of Cromer

Previously revetments, now managed retreat

Prior to the 1990s, the coast at Happisburgh was protected by revetments built in the 1950s. However, during the 1980s these revetments had fallen into a bad state of repair and storm waves were eroding the coastline at fast rates, resulting in the loss of some houses along Beach Road. In 1989, North Norfolk District Council (NNDC) identified the need for a major investment in new defences. For a variety of reasons, these were not provided, and the coastline eroded at a fast rate.

Table 15 Impacts on physical and human environments

Physical environment	Human environment
Erosion will allow continued exposure of the 6 hectares of SSSI cliffs at Happisburgh. However, Natural England sees this as a positive environmentally as it allows the study of the geology found there. Any processes that interrupt erosion can stop the exposure of fresh geological outcrops, as the previous ones get covered by vegetation and rock debris.	There will be a significant loss of commercial and residential properties as a result of the managed retreat option. By 2025, around 15 properties will have been lost, primarily along Beach Road, Happisburgh. In addition there will be a loss of land from the cliff top caravan park and the HM Coastguard Rescue facility. However, at this stage, no loss of cliff top heritage sites is expected.
Elsewhere in the sediment cell, the sediment eroded from these cliffs will contribute to beaches and dunes further to the south, maintaining and enhancing the dunes at places like Eccles-on-Sea and Winterton-on-Sea.	By 2050, it is expected that cumulative losses of properties will be between 15 and 20. Further loss of cliff top caravan park land at Happisburgh. By this stage, the grade 1 St Mary's Church and the grade II manor house at risk of erosion.
There will be loss of grade 1 agricultural land to erosion from the sea, totally up to 45 hectares by 2105.	By 2105, total residential losses are expected to be around 20 and 35 properties and the SMP foresees the probable loss of St Mary's Church and manor house.

Sustainability

The managed retreat strategy is sustainable to a large extent. The loss of land will be a major input of sediment into the sediment cell, and it will be transported south along the coast to enhance the beaches and dunes found there. This will not only protect settlements such as Eccles-on-Sea and Winterton-on-Sea but it will also help to maintain and develop natural habitats. It will also allow the coastline to operate in a natural manner. So, despite the loss of residential properties, cultural heritage and economic assets in the north, these losses are not sufficient to justify the costs of protecting the coastline here, never mind the negative impact such interventions would have in other areas to the south.

Exam tip

You may be asked to evaluate the management for this case study. Any evaluation must look at both positives and negatives and come to a balanced conclusion.

Summary

- In England and Wales, coastal management is delivered through Shoreline Management Plans (SMPs).
- These seek to produce long-term, sustainable management strategies and explore three key factors: cost–benefit analysis (CBA), environmental impact assessments (EIAs) and technical viability.
- They have four options for each section of coast: hold the line, advance the line, managed retreat and no active intervention.
- In delivering sustainable management, a variety of hard engineering strategies (sea wall, revetments, rip-rap, gabions and groynes) and soft engineering strategies (beach nourishment, dune regeneration and managed retreat) are available, as exemplified in the case study of north Norfolk.

Option D Climate change: past and present

Natural climate change processes

Climate change processes

Long-term changes

Climate change is far from a new concept. For most of the Earth's history (about 85%), our climate was generally warmer than it is now with no permanent ice caps at either of the poles. However, there have been five major periods of ice age (defined as periods of time when there were permanent ice caps on one or both of our poles) during our planet's history:

- **Quaternary** (starting 2.58 million BP (before present), and continuing to the present day, although we are currently in a slightly warmer interglacial called the Holocene)
- **Karoo** (360–260 million years BP)
- **Andean-Saharan** (around 450 million BP)
- **Cryogenian** (around 800 million BP, when it is thought that glacial ice sheets may have covered the entire planet)
- **Huronian** (over 2 billion BP)

Around 50 million years ago, as we moved towards the Pleistocene epoch, the Earth's climate slowly transitioned from a warmer, more stable climate to a colder, more variable one, known as the Pleistocene.

Medium-term changes

During the Pleistocene period (2.6 million years to 11,700 BP), the climate was quite variable. It was characterised by longer, cold glacial periods of about 100,000 years, interspersed by shorter, warmer interglacials (interstadials) typically lasting ten(s) of thousands of years. There were 11 major glacial and many more minor glacial periods within the Pleistocene. Ice core records show that the cyclical swing between glacials and interglacials was quite rapid, occurring over the period of just a few centuries. In addition, the glacial periods had variations of temperatures within them, with several short warming and cooling periods. In contrast, the interglacials tended to have more stable temperatures, the variations then tending to be over the period of a thousand years and more subdued than the variations during glacial periods. Figure 45(a) shows changing temperatures over the past 400,000 years.

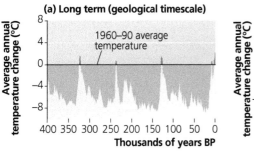

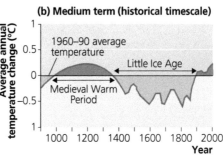

Figure 45 Changing temperatures (a) over the past 400,000 years (b) over the past 1,000 years

Knowledge check 20

Outline the long-term and medium-term climatic changes in our planet throughout its history.

For the past 11,700 years, we have been in the Holocene epoch, an interglacial period when the ice sheets have retreated back to the poles once more. During this time period, although temperatures have been more stable than they were during the glacial periods of the Pleistocene, there has been some medium-term fluctuation. From the start of the Holocene to around 5,500 years BP, temperatures were at their highest in the epoch. Since then, they have dropped slightly, albeit with some variations: between the tenth and fourteenth centuries, there was the Medieval Warm Period, followed by the Little Ice Age from the fourteenth century to the mid-nineteenth century. Since then, temperatures have been rising (Figure 45(b) — but that takes us into human climate change, a topic we will address later in this section.

Evidence for climate change

Although direct thermometer evidence for past climate data only stretch back to the mid-nineteenth century, there are other ways in which we can obtain historical data about the temperatures of the Earth.

Sources of evidence for climate change

- **Ice cores**: As ice forms in places like glaciers and Antarctica, bubbles of air are trapped. These contain samples of the atmosphere from the past. Scientists can drill down into the ice to extract ice cores and examine the air trapped in the bubbles. Ice cores have annual layers (similar to tree rings), so dating them is very straightforward. Samples up to 250,000 years old have been taken from Greenland, and up to 800,000 years old have been recovered in the Antarctic. These samples produce high resolution records of the past due to their year-by-year record and are very good records for the Holocene period.

  The balance of isotopes of oxygen and hydrogen can be analysed to determine the temperature of the Earth when the bubbles were frozen into the ice. Correlation with other ice core samples from across the world can help recreate temperatures to a high degree of accuracy.

Exam tip

Long-term climate change is measured over hundreds of thousands to millions of years, whereas medium-term is thousands to tens of thousands of years. They are to be distinguished from short-term change which is our current human-influenced change.

Knowledge check 21

Compare and contrast the past climatic records we can get from ice cores and ocean-floor deposits.

- **Pollen analysis**: All flowering plants produce pollen grains; different plants produce pollen of different shapes. Pollen is carried by the wind and deposited on surfaces. The grains can be well preserved when buried in sediments. This means that, when sediment cores are drilled out, scientists can identify the different types of plants that were growing in different time periods. From this, they are able to make estimations of the climate that existed then that supported these particular plants.
- **Ocean-floor deposits**: Sediment cores drilled from sea floors, for example from the Atlantic Ocean, can also be used to reconstruct past climates. The shells of sea creatures deposited in the sediments contain calcium carbonate. Analysis of oxygen isotopes in the shells of the sea creatures can be used to determine past temperatures (the greater the concentrations of heavy oxygen, the colder the temperatures when the creatures were alive). These sediments have allowed analysis of continuous temperature records for the entire Pleistocene period (back to around 2.6 million years BP) and so give us a longer record than ice cores. However, due to the fact that ocean floor sedimentation rates are so slow, they are of a lower temporal resolution than ice cores.

> **Exam tip**
>
> You need to be able to explain in detail how each of these sources of evidence provides information about past climates.

Causes of climate change

Astronomic: Milankovitch cycle

One characteristic of the climate during the Pleistocene epoch was the regularity of change. There were periods of time, lasting typically around 100,000 years, when the Earth was colder and extensive glaciation occurred; and these were interspersed by shorter interglacial periods (including the Holocene, our current time period).

The main explanatory factor for this pattern is astronomic and is known as the Milankovitch cycle. This describes periodic changes in the way in which the Earth orbits the Sun, and these period changes can be linked to the periodic changes in global climate during the Pleistocene. There are three main elements to the Milankovitch cycle (Table 16).

Table 16 The elements of the Milankovitch cycle

Cycle element	Description	Cycle length	Effect on Earth's climate
Eccentricity	The Earth's orbit around the Sun changes from nearly circular to more elliptical and back again.	100,000 years	This affects seasons. When the orbit is more elliptical, the seasons are more extreme: if winter occurs when the planet is further from the Sun, it will be colder; conversely, summers will be warmer.
Tilt	Currently, the axis of the Earth's rotation is tilted at 23.4°. However, this varies from 21 to 24 and back over a time period of 41,000 years.	41,000 years	The greater the angle of tilt, the more extreme the seasons. Summers will be warmer and winters colder.
Precession	Due to the gravitational pull of the Sun, the Earth 'wobbles' on its axis (in a similar way to how a spinning top wobbles as it spins).	Two cycles: 19,000 and 23,000 years	Precession determines whether summer in any given hemisphere occurs when the Earth is at its closest or furthest point from the Sun in its orbit. For example, currently, the Earth reaches its furthest point from the Sun during January. This means that the winters in the southern hemisphere are currently colder than the northern hemisphere winters, but the summers are warmer.

Milankovitch theorised that ice ages would occur when these orbital variations resulted in shorter, cooler summers in the northern hemisphere. Less of the snow that fell the previous winter melted, and that allowed a build-up of snow and ice over time. An increase in the amount of snow would result in greater albedo, and more insolation would be reflected back out into space, further prompting global cooling.

Based on the orbital variations, Milankovitch predicted that ice ages would peak every 100,000 and 41,000 years (with additional smaller peaks around every 20,000 years or so): analysis of ice cores and sea floor sediments has revealed that to be the case. The Milankovitch cycles are therefore very good at explaining the main periods of glacials and interglacials during an ice age.

Exam tip

The three cycle elements in Table 16 interact to create the overall Milankovitch cycle.

Solar

However, within these ice age cycles, there are also periods of heating and cooling over shorter timescales. One reason for this is the variation in the output of solar energy from the Sun. There are two main observable cycles:

- **Sunspots** on the surface of the Sun vary on an 11-year cycle. However, it is estimated that the difference in solar output during this cycle is about 0.1% and may only result in Earth temperature differences of 0.06 degrees.
- Nevertheless, observations reveal that sunspot activity also varies on longer timescales. For example, between 1645 and 1715, there were very few sunspots observed (known as the Maunder Minimum). This corresponded with the Little Ice Age and it may have been a contributing factor to the colder conditions.

Volcanic

Atmospheric composition can be affected by volcanic eruptions. The impact this has on global climate depends on a variety of factors, including the amount of dust and gases (carbon dioxide and sulphur dioxide) ejected into the stratosphere, and the latitude of the volcano. If dust and gases are erupted into the stratosphere, they are able to increase the reflection of incoming insolation and reduce global temperatures. For example, the Mt Pinatubo eruption of 1991 reduced global temperatures by about 0.2°C for a few years. If the eruptions occur at higher latitudes, the dust tends to get trapped in a circulation around the poles, reducing the global impact of the eruption on climate.

There is some evidence that volcanic activity can contribute to cooling on longer time-scales also. For example, the major eruptions of 1750–70 and 1810–35 may well have contributed to the cooling associated with the Little Ice Age. Further back in time, the Toba supervolcano of 75,000 years ago may have reduced global temperatures by up to 5°C for between 6 and 10 years. This may have been enough to trigger an ice age lasting 1,000 years.

Volcanic eruptions can contribute to climate change in other ways, being part of the carbon cycle. The lithosphere and mantle are part of the carbon store, and when volcanoes erupt they can release some of this stored carbon dioxide back into the atmosphere. That said, the scale of emissions is very low: total annual volcanic emissions of carbon dioxide are equal to only 3 to 5 days of carbon dioxide produced by human activities.

Continental drift

Ice ages have affected the Earth for only around 15% of our history. So there must be other factors at work that bring about the conditions for the development of these ice ages; one of the main ones is continental drift. The cooling of the global temperatures over the 5 million years into the start of the Pleistocene occurred over timescales too long to be explained by the Milankovitch cycle. Rather, it is thought that this gradual cooling resulted from the movement of continents across the surface of the Earth as a result of plate tectonics.

Around 60 million years ago, the supercontinent of Pangea began to break apart. This affected the global climate in various ways, including:

■ The redistribution of continents across the globe changed the global atmospheric and ocean circulations. As North America and South America moved away from Europe and Africa (forming the Atlantic Ocean), this caused the Arctic Ocean to become a small, almost land-locked basin. This restricted the flows of warmer water from the south and started to produce cooling around the North Pole.

■ During this time period, the Antarctic continent moved over the South Pole. Because it was not connected at all to any other land mass, it became thermally isolated and less heat was transferred poleward from the warmer latitudes. This allowed ice caps to build up on the continent around 35 million years BP and moved us closer to the Pleistocene and the ice ages that accompanied it.

■ Orogeny (mountain building): while many of the continents were moving apart during this time period, the sub-continent of India collided with the rest of Asia, forming the Himalayas and the Tibetan Plateau. This not only changed global atmospheric circulations — contributing to global cooling — but also increased global rates of weathering (during the process of weathering, carbon dioxide is extracted from the atmosphere to form bicarbonates; these then can be carried away by rivers and deposited on sea floors, where it is stored). The reduction of the greenhouse gas of carbon dioxide increases global cooling.

The end of the last glaciation and the arrival of the Holocene

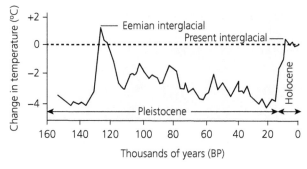

Figure 46 Temperature change from the end of the last glaciation and into the Holocene

> **Exam tip**
>
> You need to be able to make references to places when discussing the arrival of the Holocene. Look out for the references in the text.

The last glacial maximum of the Pleistocene was around 20,000 years ago. Following this, the Earth began to warm as we transitioned into the interglacial period known as the Holocene (our current epoch) (Figure 46). The warming was relatively quick when measured in geological timescales. This warming was not steady, however, and was interspersed by cooler periods when the ice sheets advanced again (for example the Younger Dryas period from 12,900 to 11,700 BP). In the UK, the name given to the more recent period of glaciation that was coming to an end then was the Devensian glaciation. During this time, sea levels were much lower, and the British Isles were joined with mainland Europe. As temperatures rose, so did sea levels. But even in 9000 BP, the Thames and the Rhine were tributaries of a larger river that flowed into the early English Channel, and the Dogger Hills was an upland area in what is now the North Sea.

The Holocene period began around 11,700 BP; the ice sheets continued to melt until around 7000 BP and temperatures continued to rise until around 5000 BP. Since then, they have fallen slightly, only to rise during the past century or so due to anthropomorphic climate change. Despite some rises and falls in the temperatures during the Holocene (for example the Medieval Warm Period and the Little Ice Age — during which there were ice fairs on the Thames), overall the climate of the Holocene has been a lot more steady than it was during the Pleistocene.

Although the complete mechanisms bringing the last glacial period to an end are not yet fully understood, current scientific thinking suggests the following set of processes are most likely to have occurred:

- Around 20,000 years ago, the Earth began to move to a period on the Milankovitch cycle where the northern hemisphere was experiencing more insolation from the Sun.
- This caused the major ice sheets over North America and Europe to begin to melt and retreat.
- Meltwater from the North American ice sheet began to flow into the North Atlantic Ocean. This addition of fresh water into the ocean reduced the density of the water, affecting the ocean currents. The North Atlantic Drift current was switched off, causing winter temperatures in Europe to fall. This was enough to cause the ice sheets to advance once again during the Younger Dryas period (12,900 BP to 11,700 BP).
- The warmer water in the Atlantic Ocean instead flowed towards the south. This resulted in changes in ocean currents in the southern Atlantic and seems to have caused carbon dioxide stored in deep ocean reserves to be brought to the surface. This addition of carbon dioxide into the atmosphere enhanced the greenhouse effect, causing global temperatures to rise quickly to the Holocene Maximum around 5000 BP (evidence from tree rings suggests that temperatures then were, on average, 2–3 degrees warmer than today's global averages).
- Since then, the relative overall stability of the Holocene climate has nevertheless experienced some fluctuations, including the Little Ice Age. These fluctuations may have been influenced by factors such as a series of significant volcanic eruptions of 1750–70 and 1810–35.

Exam tip

Make sure you use good geographical terms in your answer, including the time periods mentioned here: Pleistocene and Holocene.

Exam tip

Note here how these factors interact together to explain the rise in global temperatures: the Milankovitch cycle triggered glacial melting, which in turn affected the ocean currents, which in turn influenced the amount of carbon dioxide in the atmosphere.

Summary

- The Earth's climate has changed and varied naturally throughout its history.
- Evidence for this change can be seen in ice cores, pollen analysis and the study of ocean-floor deposits.
- The causes of natural climate change include astronomic (the Milankovitch cycle), solar, volcanic and continental drift.
- From around 11,700 years ago, the Earth's climate began to transition out of a glacial period into our current interglacial, known as the Holocene.

Lowland glacial landscapes
Glacial processes
The formation of glacial ice sheets

Glacial ice sheets are formed as follows:

- Global temperatures fall low enough to allow the accumulation of snow. This is mostly due to cooler, shorter summers. The snow that fell the previous winter does not all melt. The next winter's snowfall builds up on top of this. Year on year, an accumulation of snow builds up. This accumulation can often occur in amphitheatres or bowls in mountainous areas known as **cirques** (or corries).
- When the snow first falls, it traps some air within it. It initially becomes compressed by the weight of more snow on top of it, taking on a dense form known as a **firn**.
- If some summer melting of the snow occurs, water will percolate down into the gaps in the firn, replacing the air. This then freezes the following winter. And so the density of the mass increases as it slowly turns into solid **ice** (this can take around 20–40 years in a climate like the Alps, and a few hundred years in colder climates like the Antarctic, where there is less summer melting).
- Once the ice is formed, it is now dense enough to begin to flow downhill under its own weight. Glaciers can then flow out of the cirque and move downhill forming valley glaciers.
- These ice sheets can grow and move to cover extensive areas as a glacial period takes hold.

Glacial and fluvioglacial processes

Glacial erosion

- **Plucking** (Figure 47a): the ice freezes onto rock outcrops which, if they have been previously loosened, can be eroded. This process tends to leave behind a more jagged landscape.
- **Abrasion**: this is scouring of the valley floor and sides by angular sediment being transported by the glacier. It tends to leave behind a more smoothed landscape.
- **Frost shattering** (Figure 47a): where temperatures rise and fall above and below freezing, the rocks on the valley sides can be eroded by freeze–thaw weathering. This material falls down onto the glacier below and is incorporated into the main body of the glacier as **lateral moraine**. This can then contribute to the sediment that the glacier uses to conduct abrasion.
- **Rotational movement** (Figure 47a): as the glacier forms in the amphitheatre of the cirque, its motion rotates around a pivot point. This increases pressure on the base of the cirque, over-deepening it by erosion.
- **Extending and compressional forces** (Figure 47b): glaciers often flow down pre-existing river valleys. Where there are sections of the valley where gradient is less, the velocity of glacier movement reduces and there is compression of the glacier. This increases its thickness and puts more pressure on its base, increasing erosion. (Conversely, where gradients are higher, the glacier extends and thins, reducing erosion.)

Knowledge check 22

Explain how snow becomes glacial ice.

Exam tip

Glacial processes and landforms are those directly associated with the glacier itself. Fluvioglacial processes and landforms are connected with glacial meltwater streams (fluvial refers to rivers).

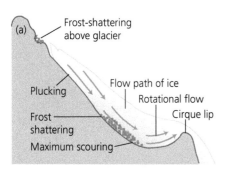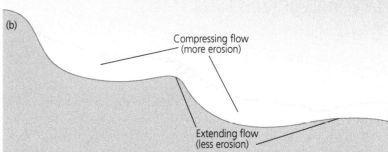

Figure 47 The processes of glacial erosion

Factors affecting the rate of glacial erosion include:

- **Gradient and velocity**: the steeper the gradient, the faster the movement and the greater the erosion rates.
- **Topography**: if a valley narrows, the ice thickens and erosion rates increase.
- **Rock type**: softer rocks or rocks with more joints erode more quickly.
- **Where two glacier tributaries join**: the thickness of the ice increases and erosion rates increase.

Fluvioglacial erosion

In addition to the direct erosion from the glacier itself, erosion may occur due to **fluvioglacial erosion processes**. Some glacial ice may melt (for example if temperatures rise sufficiently during the summer) and this water will flow as streams within and below the glacier. This moving water erodes via the usual river processes, especially via hydraulic action and corrasion.

Glacial transportation

There are three main ways that glaciers transport sediment (Figure 48):

- **Supraglacial debris** is sediment being transported on the surface of the glacier, and may be seen as darker streaks of lateral or median moraine (e.g. rock fragments loosened from the valley walls by frost shattering and falling down onto the surface of the glacier).
- **Englacial debris** is material carried within the main body of the glacier. (It can consist of supraglacial debris that has been buried by snowfalls.)
- **Subglacial debris** is sediment carried along at the base of the glacier, either in the ice itself or in meltwater streams.

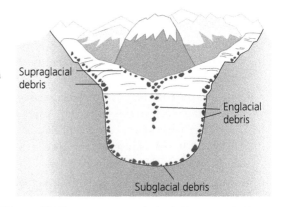

Fluvioglacial transportation

Figure 48 The processes of glacial transportation

In addition, sediment may be transported by **fluvioglacial transportation processes** via meltwater streams by means of the usual river processes: solution, suspension, saltation and traction.

> **Exam tip**
>
> Top-level answers always make very good used of terminology. Learn and use all these erosional terms in your answers in the exam.

> **Exam tip**
>
> Find out more about the fluvial (river) processes of erosion and transportation in the AS1 Student Guide in this series.

Glacial and fluvioglacial deposition

Drift is the overall term for all forms of deposits left behind by a period of glaciation. Table 17 identifies the differences between till and fluvioglacial deposition.

Table 17 Glacial till and fluvioglacial deposition

Glacial till	Fluvioglacial deposits
■ Left behind by the glacier as it retreats ■ Unsorted, i.e. consists of sediment of the full range of particle sizes ■ Angular ■ Long axes orientated in the direction of glacial movement	■ Deposited by meltwater streams either during the glaciation or by the re-deposition of glacial till subsequently moved by water ■ Sorted, i.e. particles of similar size deposited together ■ More rounded (due to attrition)
Landforms include: till, drumlins, moraines, erratics	Landforms include: outwash plains, eskers

Knowledge check 23

Distinguish between the deposition of glacial till and fluvioglacial deposits.

Glacial and fluvioglacial landforms

Glacial landforms

Exam tip

Make sure you can clearly distinguish between glacial and fluvioglacial landforms.

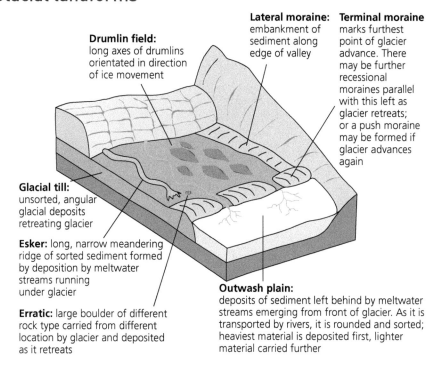

Drumlin field: long axes of drumlins orientated in direction of ice movement

Lateral moraine: embankment of sediment along edge of valley

Terminal moraine marks furthest point of glacier advance. There may be further recessional moraines parallel with this left as glacier retreats; or a push moraine may be formed if glacier advances again

Glacial till: unsorted, angular glacial deposits retreating glacier

Esker: long, narrow meandering ridge of sorted sediment formed by deposition by meltwater streams running under glacier

Erratic: large boulder of different rock type carried from different location by glacier and deposited as it retreats

Outwash plain: deposits of sediment left behind by meltwater streams emerging from front of glacier. As it is transported by rivers, it is rounded and sorted; heaviest material is deposited first, lighter material carried further

Figure 49 Glacial and fluvioglacial landforms

Till

Till is the overall term given to all sediment deposited directly by the glacier itself. Its main characteristics are that the sediment is unsorted (i.e. it consists of a mixture of the full range of sediment sizes) and angular as it has not been rounded while being transported by rivers. It is made up of the rock that the glacier passed through. The long axes of the particles in the till are aligned to the direction of the glacier movement.

Drumlins

Drumlins are oval-shaped small hills, typically 1 km long, 50 m high and 500 m wide (Figure 50). As the glacier moves, at times it erodes more material than it can transport, and deposition occurs, forming the drumlin. As the glacier advances, it streamlines the mound and gives the **lee** side a more gentle gradient than the steeper **stoss** end. The end result is a collection of drumlins (a drumlin field) whose long axes are orientated in the direction of the ice flow, their steeper slope showing the direction from which the ice approached.

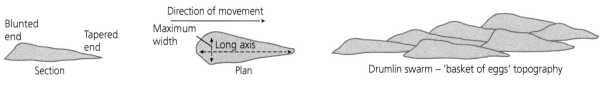

Figure 50 Drumlins

Erratics

Erratics are large boulders of rock type different from the surrounding rock, deposited on the landscape. They are eroded by the glaciers many kilometres away, transported and eventually deposited.

Moraines

Moraine is a landform produced by the deposition of sediment being transported by the glacier as it melts and retreats. It can take various forms:

- **Lateral moraine**: an embankment of sediment found along the edge of a valley previously occupied by a glacier.
- **Medial moraine**: an embankment of sediment found in the middle of a glaciated valley, formed by the merging of two sets of lateral moraine when two tributary glaciers meet.
- **Terminal or end moraine**: a linear mound (or series of mounds — the moraine may have been bisected by rivers of meltwater) of till running perpendicular to the glacier front, showing the location of the maximum extent of the glacier.
- **Recessional moraines**: similar to the terminal moraine, but these mark places where the glacier paused for a while during its retreat, allowing deposition to occur.
- **Push moraine**: this forms if a glacier, which has already formed a terminal moraine, then advances again. This pushes the terminal moraine forward, forming a steeper mound.
- **Ribbed moraine**: this moraine is found in groups and consists of lines of glacial till running perpendicular to the direction of ice flow, often forming crescent-shaped ridges. The ridges are asymmetrical: the up-ice side is more gently sloping and the down-ice side steeper. Their formation is uncertain. One theory suggests they form in sediment-rich glaciers, with the ridges being formed by deposition and reshaping when the glacier experiences compression; another suggests that they may be formed by the reshaping of marginal moraines by a 90 degree change in the direction of ice movement.

Exam tip

Practise drawing properly formed and well-annotated diagrams to show the characteristics of a drumlin.

Knowledge check 24

Distinguish between terminal, recessional and push moraines.

Fluvioglacial landforms

Eskers

Eskers are long, narrow, meandering ridges of sorted sediment. They are formed by deposition of sediment by fluvioglacial meltwater streams running through meltwater tunnels under a glacier. They are then revealed as the glacier retreats.

Outwash plains

Outwash plains are areas of fluvioglacial deposition left behind by meltwater streams that emerge from the front of the glacier. As the material is being carried by river water, it is sorted: the largest sand particles are deposited first, followed by silts, and the clay-sized particles are carried furthest. The material may come from sediment carried in the glacier itself, or it may be made up of some of the glacial till that is eroded by the meltwater streams as they emerge from the glacier, before being transported further away and ultimately being deposited.

> **Exam tip**
>
> One common feature of fluvioglacial landforms as opposed to glacial landforms is that the sediment is sorted by size. This is because it is being transported by rivers. Refer back to the Hjulstrom curve from the AS1 Student Guide in this series for more information on this.

Case study

Lowland glacial region in southeast Northern Ireland

Glacial and fluvioglacial features in Co. Down

Many of the glacial and fluvioglacial features in Co. Down were formed during the Killard Point Stadial (14,500 to 13,000 BP), the last major Northern Irish glacial period before the Holocene. Much of the ice that affected Co. Down flowed out from the Lough Neagh ice sheet, and moved SSE across the county.

Esker ridge at Lisburn is a 2.5 km long ridge from Dunmurry to Lisburn. Causeway End Esker is a 60 m ridge that runs from Brookmount to the western edge of the city of Lisburn. It is made up both of local sediment, e.g. Triassic sandstone, but also from further afield, e.g. Tardree in Co. Antrim. These mark meltwater channels below the glacier and ultimately feed sediment into the Malone Delta.

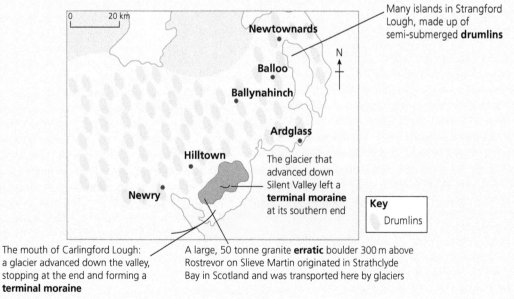

Many islands in Strangford Lough, made up of semi-submerged **drumlins**

The glacier that advanced down Silent Valley left a **terminal moraine** at its southern end

A large, 50 tonne granite **erratic** boulder 300 m above Rostrevor on Slieve Martin originated in Strathclyde Bay in Scotland and was transported here by glaciers

The mouth of Carlingford Lough: a glacier advanced down the valley, stopping at the end and forming a **terminal moraine**

Key
Drumlins

Figure 51 The glacial and fluvioglacial features of Co. Down

The extensive **drumlin field** in Co. Down covers 1,600 km² and consists of around 3,900 drumlins. It is part of a more extensive field that runs as far west as Co. Sligo and as far south as Co. Cavan and Co. Monaghan. The drumlins have a NNW/SSE trend, showing the direction of movement of the ice from the Lough Neagh ice sheet. They range in length (180–850 m) and width (1–350 m) but on average they are 350 m long and 210 m wide. They are up to 50 m high.

There are various local examples of moraines in Co. Down which mark the end of the ice sheet.

The benefits and problems of socioeconomic development

The city of Newry is located at the head of Carlingford Lough, which was carved out during the last glacial period and onto which drumlins were deposited. It is placed strategically on the junction of the A1 Eastern Seaboard Key Transport Corridor and the A28 Link Corridor as well as on the Belfast–Dublin railway line and, as such, has many economic opportunities. The Banbridge/Newry and Mourne Strategic Development Plan (2015) recognises this and commits to developing its role as an employment location by releasing 124 hectares of land for economic development including light industry and business.

That same plan allows for the growth in the number of houses available in the city, allocating 4,655 housing units across 130 hectares of land. However, it also recognises the constraints on the outward expansion of the city, including 'topographical factors' such as the 'elevated land associated with the town's drumlin landscape'. Similarly, the expansion of the town of Ballynahinch is limited by the drumlins surrounding it. The Strangford and Down Strategic Development Plan (2015) highlights various landscape features such as the drumlins to the west beside Grove Road and to the east of the town. It sees them as forming part of the 'landscape setting' of the town in the Northern Ireland Landscape Character Assessment Report. As such, they act as both a constraint to urban expansion and also an environmental asset to the town itself. And, towards the north of the county, the drumlins also act to constrain urban expansion. For example, the prominent drumlin at Bowtown Road at Newtownards acts as a good visual stop to development.

There are many smaller settlements within the area too. For example, the village of Hilltown is situated on a drumlin close to the Mourne Mountains. Since 1986, it has been part of the Mourne Area of Outstanding Natural Beauty (AONB) so the Strategic Development Plan (2015) seeks to protect this important lowland post-glacial natural resource by tackling urban expansion into the surrounding rural area and preventing ribbon development into the countryside. In fact, the drumlins help some of the smaller villages such as Balloo (which acts as an important local service centre with its range of shops, including a pub, a restaurant, a post office, a garage, two churches and two community halls and a vet) to be well integrated into the landscape, as they hide these villages within the landscape.

> **Exam tip**
>
> Produce a summary table of this part of the case study, classifying the notes as 'benefits' and 'problems'.

Farming is an important activity in the lowland area here. In 2015, about 30% of all the farms in Northern Ireland were located in this region. The drumlin landscape provides some benefits including soils that are reasonably well drained on the drumlins themselves, and farming here is easier than it is in much more mountainous areas. However, there are challenges of farming on this landscape too. For example, the soils are thin and, due to the unsorted nature of glacial till, they have a lot of rocks in them. This means that, in order to plough the land, it is often necessary to use rock breakers. The steepness of the slope angles on some drumlins is challenging for tractors. The drumlin hollows are characterised by poor drainage all year round and so are either not part of the productive farmland or require expensive drainage. Given the NNW/SSE orientation of the drumlins, the more northerly facing side gets less sunlight and is less well drained than the other →

side, so it is less useful for arable farming. As a result, 95% of farming here is livestock.

The flooded lowland area of Strangford Lough AONB, including its many drumlin islands, provides many economic and recreational opportunities. There are 12 sailing and boat clubs here (more than the rest of NI combined). The Ardglass Marina has 80 berths for both local and visiting yachts. There are other recreational activities in Co. Down. For instance, since the Giro d'Italia event in Northern Ireland in 2014, the annual Gran Fondo 175 km cycle race through the drumlins of Co. Down has become very popular.

Summary

- Ice sheets begin to grow as the snow from winter does not completely melt during the following summer and begins to turn into ice over time.
- Various processes contribute to glacial erosion, including: plucking, abrasion, frost shattering, extending and compressional forces, and rotational movement.
- In addition, erosion may occur as a result of fluvioglacial processes caused by glacial meltwaters.
- Glaciers transport sediment as supraglacial debris, englacial debris or subglacial debris, as well as the usual fluvial transportational processes of solution, suspension, saltation and traction.
- Glacial deposits can take the form of till or fluvioglacial deposits.
- These processes create various glacial and fluvioglacial landforms such as till, drumlins, erratics, moraines (lateral, medial, terminal, recessional, push or ribbed), eskers and outwash plains.

Current global climate change: human causes and impacts

Evidence for short-term climate change and links to air pollution

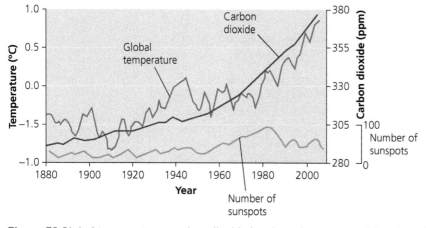

Figure 52 Global temperature, carbon dioxide levels and sunspot activity since 1880

There is widespread scientific consensus that the Earth has been experiencing warming over the past decades. The 2014 report by the Intergovernmental Panel on Climate Change (IPCC) states that: 'Warming of the climate system is unequivocal and, since the 1950s, many of the observed changes are unprecedented over decades to millennia. The atmosphere and ocean have warmed, the amounts of snow and ice have diminished, and sea level has risen … Each of the last three decades has been successively warmer at the Earth's surface than any preceding decade since 1850. The period from 1983 to 2012 was likely the warmest 30-year period of the last 1400 years in the northern hemisphere.'

Recent short-term climate change

- **Rises in observed air temperatures**: temperature records dating back to the start of the twentieth century indicate that the average temperature of the Earth's surface has increased by about 0.8°C in the last 100 years. About 0.6°C of this warming occurred in the last three decades. The temperature rises have seen regional variations: in the USA, the average temperature between 1990 and 2008 was around 0.5°C warmer than average between 1940 and 1980. However, during the same time periods for the Arctic, the average temperature was closer to 2°C warmer.
- **Ocean warming**: 90% of the additional energy on Earth due to warming has been stored in the oceans. This causes thermal expansion of the water, raising sea levels — levels have risen by around 20 cm since 1900.
- **Loss of Arctic sea ice**: the average coverage has decreased by about 4% per decade between 1979 and 2012. In 2012, the ice extent reached a record minimum that was 50% lower than the 1979–2000 average.
- **Loss of ice sheets**: there has been a loss of ice sheet mass from both the Antarctic and Greenland ice sheets between 1992 and 2011.
- **Glacier retreat**: glaciers grow or retreat due to natural factors, but analysis of average movement over decades (decadal averages) reveals that around the globe the glaciers retreated during the 1940s, then were stable or growing until the 1970s, but since the mid-1980s, most glaciers have been in retreat. In France, all six of the main glaciers in the country are in retreat.

Causes: the link to air pollution

We have seen that some factors can cause short-term climate change, most notably changes in solar output.

However, observed data of solar output show that, while temperature trends have been rising noticeably since the 1980s, the trends of solar energy outputs have been falling. At the same time, carbon dioxide levels have been rising. In fact, the IPCC 2014 report stated that scientists are 95% certain that the bulk of this warming is due to human factors, especially the burning of fossil fuels increasing the amount of carbon dioxide in the atmosphere. In 2013, recorded carbon dioxide levels in the atmosphere surpassed 400 ppm (parts per million), the first time levels have been that high in 4.5 million years. The biggest emitters of carbon dioxide are China (over 35,000,000 kt in 2014) and the USA (over 10,000,000 kt in 2014).

Greenhouse gas emissions from humans have continued to rise. Between 1970 and 2000, they rose by 1.3% per year; they increased by 2.2% per year from 2000 to 2010. This has been driven mostly by population growth and increasing economic development. Around three-quarters of human emissions of carbon dioxide result

Exam tip

You should be able to refer to places to illustrate your answer as you discuss evidence for short-term climate change.

from burning of fossil fuels; the rest is mostly the result of cutting down carbon-absorbing forests in places like Brazil. Since the Industrial Revolution began in 1750, carbon dioxide levels have risen by more than 30% and methane levels have risen more than 140%.

The addition of these greenhouse gases enhances the naturally occurring greenhouse effect. Solar radiation arrives at the Earth at shorter wavelengths and more easily passes through the greenhouse gases in the atmosphere, heating the Earth. The Earth re-radiates this, but at longer wavelengths. These are trapped by the greenhouse gases in the atmosphere, allowing less heat to escape and causing a warming of our atmosphere.

Present and potential impacts of climate change on MEDCs and LEDCs

There have been a number of direct impacts on our planet resulting from climate change in recent decades. These include the loss of sea ice, ice sheets and glacier retreat as discussed above. In addition to these, the following impacts have been observed (Figure 53).

> **Exam tip**
>
> Short-term climate change over the past century or so is mostly anthropomorphic (i.e. caused mainly by human actions). Make sure you can make the link between rising temperatures and increasing air pollution since the Industrial Revolution, especially carbon dioxide.

In **North America**, the tundra in **Alaska** is already melting. Soil temperatures at 1 m depth have risen by 4°C and at 20 m by 2.5°C. The warmer, drier summers experienced there are increasing the wildfire risk. There were more large fires in Alaska in the 2000s than since records began in the 1940s.

In recent decades, there has been an increase in more extreme weather events in **Europe**. For example, in 2007, England and Wales experienced the wettest July on record since 1766, producing £3 billion of damage. In 2011, France had its hottest summer since 1880 and grain harvests dropped by 12% that year. Across the entire continent in 2003, it was the hottest summer in at least 500 years, resulting in an estimated 70,000 heat-related deaths.

Mountainous areas in **Europe** will experience glacier retreat, reduced snow cover and winter tourism, and extensive species losses (in some areas up to 60% under high emissions scenarios by 2080). In **southern Europe**, climate change is projected to lead to higher temperatures and more drought, and to reduce water availability, hydropower potential, summer tourism and crop productivity. Warming temperatures will increase the health risks due to heatwaves and the frequency of wildfires.

▪ Present impacts

▪ Potential impacts

Warming in the **Rockies** is projected to reduce the snowpack, leading to more winter flooding and reduced summer flows. In the first half of the twenty-first century, moderate climate change is projected to increase aggregate yields of rain-fed agriculture by 5–20% overall (but with important variability among regions within North America). Cities that currently experience heatwaves are expected to experience more, longer and hotter heatwaves during the course of the century, with negative health impacts. Warming in **Alaska** will result in thawing of the permafrost and lead to an increased threat from wildfires.

In **southeast Asia**, coastal cities will be under intense stress due to climate change. A sea-level rise of 30 cm, possible by 2040 if current rises in carbon dioxide continue, would cause massive flooding in cities and bring saltwater pollution to low-lying cropland. **Vietnam's Mekong Delta**, a global rice producer, is particularly vulnerable to sea-level rise. A 30 cm sea-level rise there could result in the loss of about 11% of crop production. At the same time, storm intensity is likely to increase.

Sea level rise is already impacting low-lying small islands in the Pacific. During the twentieth century, sea levels rose by an average of 1.8 mm per year. From 1993 to 2003, this increased to 3.1 mm per year. Rising sea levels (along with increased frequency of storms) is contaminating the fresh water stored in the soil in **Kiribati.**

In **sub-Saharan Africa**, food security will be a major challenge from climate change, with dangers from droughts, flooding and shifts in rainfall. Between 1.5°C and 2°C warming, drought and aridity will contribute to farmers losing 40–80% of cropland conducive to growing maize, millet and sorghum by the 2030s–2040s. In a 4°C warmer world, around the 2080s, annual precipitation may decrease by up to 30% in southern Africa, while **east Africa** will have higher rainfall totals. Ecosystem changes could reduce food for grazing cattle.

From summer 2011 to summer 2012, a severe drought (the worst in 60 years) affected the entire **east Africa** region. There have been challenges to water supply in Africa. For example, the glaciers on Mt Kilimanjaro have been retreating (around 80% loss since 1912), reducing water supply to several major rivers. Rainfall totals have been affected: 2010–11 was the driest year since 1950–51, causing challenges to food supply.

There will likely be significant variability in water supply across **south Asia** as a result of climate change. Inconsistencies in the monsoon season and unusual heat extremes will affect crops. Loss of snowmelt from the Himalayas will reduce the flow of water into the **Indus, Ganges and Brahmaputra** basins, leaving hundreds of millions of people without enough water, food, or access to reliable energy. **Bangladesh** and the Indian cities of Kolkata and Mumbai will be confronted with increased flooding, intense cyclones, sea-level rise, and warming temperatures.

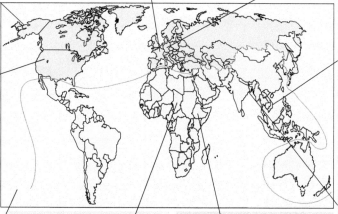

Figure 53 The present and potential impacts of climate change on MEDCs and LEDCs

Exam tip

Again, you need to be able to refer to places and examples to illustrate the impacts. In addition to those mentioned here, see the Kiribati case study on pp. 63–64 for much more detail on the impacts of sea level rise on this island nation. The AS1 Student Guide in this series has a case study on the impacts of climate change in Alaska.

Exam tip

Although it is not possible to directly link one specific extreme weather event to climate change, an increase in extreme weather events in general is nevertheless to be expected in a warming world where climate is changing.

Summary

- There is widespread scientific agreement that the recent rises in global temperatures over the past 100 years or so have been driven primarily by human actions.
- These specifically include pollution of the atmosphere by greenhouse gases, including carbon dioxide.
- While it is difficult to tie one extreme weather event to climate change, there are already many examples of the kinds of climatic changes we could expect to see occurring in a warmer world. And into the future, these impacts are projected to increase in both MEDCs and LEDCs.

Managing global climate change

Attempts to address global climate change

There is now widespread agreement of the reality of human-induced climate change. Much of the debate therefore currently revolves around how we should respond to the threats and challenges it presents. The responses fall into two broad categories: mitigation and adaptation.

Exam tip

Be clear in your own mind of the difference between mitigation (trying to reduce the causes of climate change) and adaptation (responding to the impacts of climate change).

Mitigation

Mitigation is a human intervention to cap and eventually reduce the amount of greenhouse gases in the atmosphere by reducing the sources or developing the sinks (stores) of greenhouse gases.

Carbon capture

The aim of carbon capture is to collect carbon dioxide from sources of emission (such as power stations), then transport it elsewhere where it can be put into terrestrial stores or sinks so that it remains there rather than entering the atmosphere. The carbon dioxide is dissolved in water and injected into deep rock stores where it reacts with the rock to form carbonates.

One report suggests that there is storage capacity in North America for 900 years of carbon storage, based on current usage levels. In the UK, the depleted oil and natural gas fields in the North Sea could be used for carbon storage. In 2016 in Iceland, new technological approaches led to the solid carbonates forming in two years (compared to the hundreds of years that had previously been predicted). This has led the Icelandic government to aim to bury 10,000t of carbon dioxide per year.

However, there remain concerns about the possibility that the carbon dioxide may leak back out into the atmosphere, perhaps in the more worrying form of methane. In addition, there are considerable extra energy costs involved in carrying out the capturing of carbon. For example, for coal-based power stations, the extra power needed ranges from 24% to 40%, significantly increasing the amount of fuel needed to produce the same amount of power.

Reducing greenhouse gas emissions

There are various ways to reduce the amount of greenhouse gases going into the atmosphere in the first place.

- Reduce emissions from burning fossil fuels by:
  - Using renewable energy sources including: wind, HEP, solar and wave energy and/or making energy production more efficient, e.g. by reducing dependence on burning coal (coal emits about twice the amount of carbon dioxide than natural gas when burned). Much of China's energy supply for its economic development has come from the burning of high-emitting coal. However, in recent years, China has been moving away from the use of coal. In 2015, its coal production and coal-fired energy production both fell by 3% and the government banned the development of new coal mines for three years. At the same time, low-carbon energy production sources increased by more than 20% in China.
  - Using alternative power sources for transportation, such as electric cars, congestion charging, incentivising increased public transport use.
- Changing land use practices: the burning of forest (especially the tropical rainforests) to clear land for agricultural purposes contributes to between 12% and 30% of human emissions of greenhouse gases. Initiatives such as Reducing Emissions from Deforestation and Forest Degradation (REDD) seeks to incentivise governments in tropical LEDCs to act to protect forest areas and thus reduce this. One country involved with REDD is Argentina. Between 1996 and 2011, it lost around 15% of its forested area. The country became a REDD partner in 2009 and is engaged in working with the UN to reduce deforestation in its forests.

> **Exam tip**
>
> One of the ways you can get the kind of depth and detail required for A2 Geography is to be able to refer to the places and examples given here of the mitigation and adaptation strategies. In fact, questions may explicitly require such references, so learn the places referenced in the text.

Adaptation

Reducing vulnerability

Even if the mitigation strategies manage to curb the emission of greenhouse gases, there is a certain amount of climate change that will still occur. This means that countries will need to take actions to adapt to a changing climate; this is especially the case for LEDCs as not only are they often less resilient to natural disasters, they are also expected to bear the brunt of the impacts of climate change. The IPCC defines vulnerability as being determined by three factors: exposure to hazards (such as sea level rise); sensitivity to those hazards (such as coastal farms on low lying land); and the capacity to adapt to the hazards (such as whether the farmers have the money to invest in salt-water resistant strains of crops).

Adaptation can take many forms:

- **Resilient agricultural practices**: climate change will lead to changes in global rainfall patterns. With around 80% of global agriculture reliant on rain for water, this could have serious impacts, especially for the 850 million poorest people in Africa and Asia. Adaptation strategies include using drought resistant crops, irrigation programmes that promote sustainable use of groundwater supplies, and the use of water storage systems for drought prone areas (for example, in Niger, the use of small basins to collect rainwater has led to a three- to four-fold increase in crop yield).
- **Glacial flood prevention**: the melting of glaciers is leading to a much increased risk of glacial meltwater flooding. To address this, terminal moraines can be replaced with concrete dams.
- **More resilience in the face of extreme weather events**: a warmer world is a more energetic world and one in which extreme weather events are more likely. The UK is seeking to prepare for an increased flood risk, for example, by investing in river and coastal flood defences.
- **Adapting to sea level rise**: with half the world's population living in coastal areas, and around 10% living in areas particularly at risk to sea level rise due to climate change, adaptation here is very important. Bangladesh has, with assistance from the World Bank, been using a variety of strategies including coastal embankments and salt-water resistant crops which are more resilient to sea water flooding. Around 1 million at risk people in Bangladesh have already been assisted.
- **Financial assistance for LEDCs:** LEDCs are, at the same time, most at risk from climate change and have the least capacity to adapt to it. As a result, there have been various attempts to help provide funds so that LEDCs can begin to adapt to climate change. For example, in 2000, an Adaptation Fund was set up of $1 billion a year for LEDCs; as part of the Paris Agreement in 2015, MEDCs agreed to provide $800 million per year by 2020 to support the Least Developed Countries in their adaptation efforts.

Knowledge check 25

Distinguish between mitigation and adaptation.

Evaluation of the progress of international action on climate change

Despite the broad consensus on the reality of climate change resulting primarily from an enhanced greenhouse effect produced by rising human-induced carbon dioxide emissions, global emissions of carbon dioxide have continued to rise over the past few decades. This is despite efforts made to reduce emissions. In this section, we will evaluate two of those measures: the Kyoto Protocol and the IPCC.

The Kyoto Protocol

- The Kyoto Protocol was agreed in an international conference in Japan in 1997. In total, 40 MEDCs signed up to making cuts in carbon dioxide emissions as part of the Kyoto Protocol, including: Europe, North America, Japan, Australia and New Zealand and the 'Economies in Transition' (mostly central and eastern European countries formerly part of the Soviet Union).
- These countries committed to making on average 5% cuts in emissions below their 1990 levels.

- LEDCs (including China) were not required to make cuts in emissions under the Kyoto Protocol. This was part of the 'common but differentiated responsibility' approach of the Protocol.
- In 2010, the Kyoto Protocol was extended to 2020, with a replacement to be negotiated by 2015 to come into effect in 2020 (the Paris Agreement was signed in 2015) – although only 23 countries have signed up to this extension.

Evaluation

Positives

- Some countries that signed up to make cuts in emissions managed to do so, including many of the EU countries. For example, the UK had cut its carbon dioxide emissions to 35% below its 1990 levels by 2014.
- It sought to minimise costs to the countries making cuts by including flexibility mechanisms. These included carbon trading. Governments allocated emission permits to companies. Those companies that struggled to make cuts due to costs could buy permits from those companies more able to make cuts. A similar arrangement existed internationally, as countries could buy emission allowances from countries that exceeded their targets. Some argue that this is a good part of Kyoto as overall cuts could be made at the lowest cost to a country.
- Some say that the Kyoto Protocol was a vital first step in international agreement and that it paved the way for subsequent agreement in Paris in 2015. This agreement is much more far reaching and includes all types of country, not just MEDCs.

Negatives

- The USA, the world's biggest carbon dioxide emitter per capita, refused to sign up to cuts following Kyoto, stating that it would harm their economy too significantly. In 2011, Canada withdrew from Kyoto (the government prioritised the exploitation of oil sands in Alberta instead).
- China, the world's biggest overall emitter, was not required to make cuts as part of the Kyoto Protocol.
- Data appears to show that emissions have fallen in the EU and in 'countries in transition' (former central and eastern European countries previously part of the Soviet Union).
- However, eastern European countries (new EU members in this data) recorded cuts during this time period largely due to the collapse of manufacturing industry following the breakup of the former Soviet Union. So, these are not cuts due to environmental policies as much as evidence of changing economic context. More recently, their emissions have begun to increase again.
- Western European countries have managed to reach these figures in part due to carbon trading which means companies in the EU can trade emissions permits with each other. When these traded emissions are included, Italy for example moves from being ahead of its target to substantially behind.
- China's emissions show considerable growth compared to all other countries. However, much of this growth has been driven by exports to MEDCs. So, when you look at the carbon footprint of a country that includes its imports, Europe managed to reduce its emissions by only 1% from 1990 to 2008.

- By 2012, some MEDCs were far behind their emissions targets, including New Zealand (60% behind its target) and Australia (50% behind).
- Overall, total global carbon dioxide emissions from MEDCs and LEDCs have risen since the adoption of the Kyoto Protocol.

Overall, then, the Kyoto Protocol has been very limited. Although some countries that signed up to cuts have managed to reduce their carbon dioxide emissions, many have not. Furthermore, the exclusion of China and the failure of the USA to sign up to cuts have meant that total world carbon dioxide emissions have continued to grow in the years since Kyoto. At best, it might have slowed down the growth of emissions and created a context for the Paris Agreement in 2015.

Intergovernmental Panel on Climate Change

The Intergovernmental Panel on Climate Change (IPCC) was set up in 1988. It is made up of representatives of governments from across the world and of scientists whose role it is to assess the latest science and evidence on climate change and to make recommendations on actions that should be taken. It publishes detailed Assessment Reports every few years (led by three working groups) that have three main sections:

- the latest scientific evidence on the **nature** of climate change
- the **impacts** (current and projected) of climate change
- the **mitigation and adaption strategies** that could be adopted.

Its main role, then, is to provide a robust and reliable evidence base for the nature of, impacts of and response to human-induced climate change, so that policy makers can have access to information to help them make policy decisions in their countries.

Evaluation

Positives

- The IPCC reports are robust and evidence based, and seek to draw together the best scientific evidence available in the world.
- The reports are wide ranging, covering the nature of, impacts of, and response to climate change.
- The reports represent a wide range of scientific view and opinion. This includes views that are different to the general consensus. In fact, the lead authors of the reports are required by IPCC guidelines to 'record views in the text which are scientifically or technically valid, even if they cannot be reconciled with a consensus view'. This helps its scientific and political credibility.
- The reports are based on scientific consensus, which means that they have very broad acceptance across the scientific community. The goal during the compilation of the reports is to have all of the working group's authors agree that each side of the scientific debate has been represented fairly.
- As the reports are designed to inform political decision making, government representatives participate in drawing up the reports and help reach the overall consensus. This means that governments cannot easily criticise or dismiss a report that they themselves have helped shape and approved during political negotiations.

Exam tip

Evaluations require you to look at both sides of the argument in a balanced way before coming to a conclusion. When it comes to evaluating the Kyoto Protocol, even if your overall conclusion is that it is not very effective, you still will have to examine its positive aspects as well as its limitations.

Exam tip

Do some wider reading and research on the more recent attempts to tackle climate change. In particular, explore the 2015 Paris Agreement. In what ways is it better than the Kyoto Protocol? In what ways does it still fall short of the kinds of actions needed to address climate change?

Negatives

- Owing to the conservative nature of the reports, the IPCC tends to underestimate and understate the future risks of climate change.
- The Third Assessment Report in 2001 had controversy surrounding the use of the 'hockey stick graph'.
- The IPCC does not carry out its own research but draws together scientific papers from all across the world. There is a cut-off date for submission of these papers before the reports are published. This means that evidence that comes to light after this date cannot be included. As the scientific evidence in the realm of climate change is progressing quickly, this can be seen as a limitation of the methods used to produce the reports.

Overall, the IPCC plays an invaluable role in collating evidence and recommending actions and its approach ensures that it has strong credibility and reputation.

Summary

- The main strategies being used to address the impacts of climate change are mitigation (using strategies including carbon capture and reducing greenhouse gas emissions to tackle the causes of climate change) and adaptation (taking actions to reduce people's vulnerability to the impacts of climate change).
- There have been various attempts to reach international agreement on tackling climate change. The Kyoto Protocol of 1997 was a significant first attempt to do so, but it fell far short of agreeing the kind of actions needed to address climate change. More recently, the Paris Agreement of 2015 was a wider and more far-reaching agreement, although it too has its critics.
- The IPCC is an international organisation tasked with pulling together the latest scientific research on climate change under three headings: evidence, impacts, strategies (mitigation and adaptation). It plays an important role in providing a robust scientific underpinning for political discussions about climate change, but some criticise its conservative nature.

Questions & Answers

The A2 Unit 1 Geography paper includes two questions for each of the four options. **Students answer two questions, one from each of their two chosen options.**

Each question is awarded up to 35 marks, giving a total mark out of 70.

Examination skills

As with all A-level exams there is little room for error if you want to get the best grade. Gaining a grade A* is not easy in A-level geography so you need to ensure that every mark counts.

The examination papers for A2 Unit 1, Unit 2 and Unit 3 are all 1 hour 30 minutes long. There are 70 marks available for Unit 1, which means that you get just over 1 mark for every minute to work your way through the paper. You need to make sure that you manage your time carefully — you have 45 minutes to finish answering each of the two questions. If you find that you have time left over in this exam, the chances are that you have done something wrong.

Exam technique

Students often find it difficult to break an exam question down into its component parts. On CCEA exam papers, the questions are often long and difficult to understand, so you need to work out what the question is asking before you move forward. One difference between the AS and A2 examinations is that the questions for A2 geography are set in one paper and the resources are contained in a separate booklet. You need to make sure that you refer to these resources carefully. The resources could be text (from a range of sources), maps, diagrams or photographs. You need to ensure that you sort through the information carefully and use the information from the resource **to help you** answer the question.

Command words

To break down the question properly, get into the habit of reading the question at least *three* times. When you do this it is sometimes a good idea to put a circle round any command or key words that are being used in the question.

A common mistake is failing to understand the task being set by a question. There is a huge difference between an answer asking for a discussion and one asking for an evaluation.

A full list of the main command words used in the exam can be found in the specification online.

Structure your answer carefully

Sometimes the longer questions on exam papers can prevent students from achieving maximum marks. Questions that are marked from 8 to 18 marks will be marked using three levels. Later in this section we will look at some questions and give more guidance about how you should structure your answers.

One simple approach to consider is drawing up a brief plan for your answer so that you know where it is going and how you will cover all the main aspects of the question. For example, you could draw a box to illustrate each element needed in an answer and fill each one with facts and figures to support the answer, using the marking guidance to help you work out how much time to spend on each section.

Show your depth of knowledge of a particular place/case study

The extended writing questions on the exam paper are usually focused on giving the opportunity to apply knowledge and understanding of case study material to a particular question. It is really important to show what you know here.

Examiners are looking for specific and appropriate details, facts and figures to support your case. The better you know and understand your case studies, the higher the marks you can potentially achieve.

About this section

A practice test paper with exemplar answers is provided. This will help you to understand how to construct your answers in order to achieve the highest possible marks.

Some questions are followed by brief guidance on how to approach the question (shown by the icon ⓔ). Student responses are followed by comments indicating where credit is due These are preceded by the icon ⓔ. In the weaker answers, they also point out areas for improvement, specific problems, and common errors such as lack of clarity, weak or non-existent development, irrelevance, misinterpretation of the question and mistaken meanings of terms.

■ Option A

Question 1 Plate tectonics: theory and outcomes

(a) Study the diagram, which shows the tectonic features associated with the South Sandwich Islands in the Pacific Ocean.

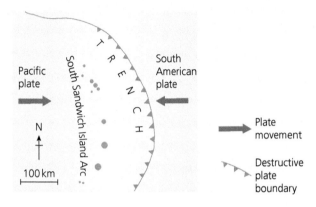

(a) Explain the formation of the landforms found at this plate margin.   (9 marks)

ⓔ **Level 3 (7–9 marks)**: Answer explains fully the process of subduction and how it forms the ocean trench and island arc system. Precise terminology is used for both processes and features.

Level 2 (4–6 marks): Question addressed, but explanation lacks depth and detail.

Level 1 (1–3 marks): Does not explain the processes in operation at this margin.

Student answer

(a) There are two landforms found here: trench and island arc. The trench is formed as the denser, heavier South American plate subducts under the slightly less dense Pacific plate. Compression forces during subduction cause the lithosphere to buckle, and friction between the two plates drags the leading edge of the non-subducting plate down, forming the ocean trench.

The processes that form the island arc start with subduction too. The oceanic lithosphere melts by subduction melting as the asthenosphere is hotter than the lithosphere. Hydration melting also occurs. Sea water brought down as the lithosphere subducts lowers its melting point. These two processes produce magma, which is less dense than the surrounding lithosphere and so it rises upwards and will form a magma chamber in the lithosphere. The magma makes its way to the surface from here, where it erupts to form a volcano. Over time, this volcano can build up to form an island above sea level. In fact, a linear arc of these islands forms parallel to the trench as multiple volcanoes erupt.

ⓔ **9/9 marks awarded** This detailed explanatory account identifies the landforms, explains the relevant processes, and makes good use of terminology.

(b) Explain how any two of the following seismic effects occur and outline their socioeconomic impacts.

Seismic shaking   Liquefaction   Tsunamis   (8 marks)

ⓔ **(2 × 4 marks):** For full marks, both cause and impacts must be addressed, showing a sound understanding, complete with effective use of appropriate terminology.

> **(b)** Seismic shaking is caused by the waves of energy that spread out from the focus of an earthquake, resulting in shaking of the ground. This shaking can be in various forms: p waves, s waves and l waves, and these cause the ground to move in complex ways, both vertically and horizontally.
>
> This shaking can have devastating impacts. In countries that do not have life safe buildings, the vertical and lateral movements can cause buildings to be shaken and twisted in shape. This can result in pancaking, where the vertical columns collapse and the concrete floors fall on top of one another, killing or trapping those unfortunate enough to be inside at the time. Pancaking was one of the main causes of death in the devastating 2010 earthquake in Haiti. In addition, other infrastructure can be damaged including pipelines, bridges, electricity pylons.
>
> Liquefaction is caused when saturated soil particles lose contact with each other during shaking, causing the soils to lose strength and integrity and act more like a liquid. After shaking, the ground re-settles and becomes more solid again.
>
> The impacts of this can cause the foundations of buildings to be compromised. Instead of making the buildings collapse and pancake, instead this can cause them to sink or topple over to odd angles. The buildings themselves may not fully collapse, but they are very dangerous and often need to be demolished and rebuilt. Cracks can appear on roads and runways during liquefaction, and sand boils can appear on the surface.

ⓔ **8/8 marks awarded** This answer clearly addresses all elements of the question: causes and impacts, and deals with two of the hazards. Good knowledge and understanding are shown, and the use of terminology is strong.

(c) Evaluate the preparation for and response management to a volcanic eruption you have studied at a small scale.   (18 marks)

ⓔ **Level 3 (13–18 marks):** Both aspects of preparation and response are dealt with in detail, with specific reference to case study material. The success and limitations of the management are clearly evaluated.

Level 2 (7–12 marks): Both aspects of preparation and response are dealt with, but in less detail. There may be an imbalance in preparations and response.

Level 1 (1–6 marks): Answers that attempt to address all aspects of the question, but with very limited detail, are confined to this level. Alternatively, answers that leave out an important element entirely (such as preparation, response, or an evaluation) would be confined to this level also.

(c) When the Pinatubo volcano first showed some signs of coming back to life in April 1991, scientists arrived to look at the activity going on. They noticed that the volcano was putting out a lot of steam and gases. They looked at where the volcano had erupted in the past. They hoped to see how it might erupt in the future from this.

As it got closer to the big eruption, the scientists noticed other things like more earthquakes. They also produced a hazards map to show what was going to happen during the eruption. They evacuated about 20,000 people living close by in June. A few days later, they evacuated lots more people. On 15 June, the volcano erupted for the first time in 75 years. Due to the evacuations, about 20,000 lives were saved.

The response to the eruption was good overall. There was a large amount of ash thrown out by the volcano covering a very big area. Around 8,000 buildings were destroyed. The government helped evacuate people and provided basic social services. They trained people in how to respond to the ash which was getting in some people's eyes and causing asthma attacks. But, on the other hand, there was disease in the evacuation centres. They monitored the threat of lahars from the ash in years to come and set up a warning system which has saved hundreds of lives. The government helped provide jobs for the people who had to leave their homes.

In conclusion, the response to the eruption was very good.

ⓔ **3/18 marks awarded** This is a poor Level 1 response that has multiple flaws. First, it is significantly lacking in depth and detail. The student shows only a superficial grasp of the case study and has not convinced that they have grasped a lot of the substance of the eruption. Second, there is limited inclusion of facts and figures. Case studies need to show the examiner a strong spatial context. This answer is too generalised, with too many vague comments. Third, it misses out an evaluation of the preparation for the eruption entirely, and the evaluation of the response is significantly imbalanced. An evaluation needs to give proper consideration to both sides of the issue.

Option B

Question 2 Tropical ecosystems: nature and sustainability

(a) Study Figure 14(c) on p. 30 which shows the climate graph for the desert location of Shuwaikh, Kuwait. With reference to the Hadley cell and the ITCZ, discuss the processes that have led to this climate. (9 marks)

ⓔ In Shuwaikh, the tropical desert is very hot (31–39°C) in the summer and temperate (15–22°C) in the winter. The total annual rainfall is 78 mm with most of the months having extremely low rainfall (March to October has less than 10 mm a month) and four summer months have no rain at all. In the summer this area is dominated by the subtropical high pressure as the air descends from the Hadley cell. In the summer the ITCZ has moved north and in the winter the ITCZ has moved much further south — allowing a little more rainfall and lower temperatures to prevail.

Level 3 (7–9 marks): The answer makes accurate reference to the temperature and rainfall figures as noted on the climate graph. The answer then makes direct reference to the role that the Hadley cell and the ITCZ plays in making the area hot and dry.

Level 2 (4–6 marks): The resource is used in relation to the reference to temperature and rainfall but the explanation of the role of the Hadley Cell and the ITCZ movement is limited.

Level 1 (1–3 marks): Answers here will show a limited understanding of the concepts of the Hadley cell and the ITCZ. There might be only a basic description of the graph.

Student answer

(a) The graph of the tropical desert in Shuwaikh clearly shows that the summer months are very hot — with a temp range of 22°C in April up to 30°C in August and back down to 18°C in October. The winter months are not as hot — fluctuating between 15°C and 18°C. There is very little rain in this place — only 78 mm each year. There is no rain between the months of June and September.

The Hadley cell helps to control air circulation around the world. The extreme heat at the equator causes the air to rise and condense. The rising air then diverges away from the equator and moves at a high level north. The air eventually sinks at around 30°C north of the equator. However, this sinking air does not bring any moisture but brings high pressure over the desert area — especially in the summer months and this means there will be a lack of rainfall in the desert areas.

Questions & Answers

ⓔ **5/9 marks awarded** This is a mid-Level 2 answer. There is some good description of the graph to show some understanding of the impact of climate in the desert. The candidate has gone on to give some useful detail about the Hadley cell that helps to explain why there is such a dry climate in the desert. However, the answer does not make any mention of the role that the ITCZ plays or the impact that the movement in the summer and winter of the ITCZ might actually play on the area.

(b) Explain the threat that large-scale development might bring to autotrophs (producers) and heterotrophs (consumers) in the trophic structure of a tropical forest ecosystem. (8 marks)

ⓔ This answer requires an explanation of the impact that large-scale development (an aspect of deforestation as caused by farming, mining or road building) might have had on the tropical forest ecosystem. The answer should focus on the role of autotrophs/plants (producers) and heterotrophs/animals (consumers) in the tropical forest.

Level 3 (6–8 marks): There is a clear and accurate explanation of the threat that large-scale development has on both autotrophs and heterotrophs in the tropical forest ecosystem. Examples and details are given from an appropriate tropical forest, e.g. the Amazon or the Monteverde Cloud Forest.

Level 2 (3–5 marks): There is some detail about the impact of producers and consumers but there might be a lack of detail or depth within the answer. Examples might be provided but these could be limited.

Level 1 (1–2 marks): Any answer in this level will not have an adequate explanation of the role that producers or consumers can play within the tropical forest. There will be a lack of depth and much of the question will remain unanswered.

(b) The trophic structure in the tropical rainforest is dominated by the massive number of trees. The hot and wet climate means that trees and vegetation will grow quickly and create a complicated, dense canopy of vegetation. The Amazon rainforest contains over 40,000 different plant species. These autotrophs are the producers that convert sunlight into food energy. The primary consumers (herbivores) in the food chain such as monkeys, frogs and parrots might eat the plant life. The primary consumers are then consumed by the secondary and tertiary consumers (carnivores) like the snakes, bats and the big cats like jaguar and puma that live in the forest.

The problem in the tropical forest is that this balance at each of the trophic levels is fragile. Commercial logging and industrial approaches to farming in the Amazon have led to an increase in tree clearance and deforestation. This means that some species will die out as their main source of food has disappeared. As one animal becomes extinct this puts pressure on the other animals further up the trophic hierarchy — as food sources are removed, the amount of energy available will be reduced and larger species of animal will also begin to die out as they don't get the food that they require.

ⓔ 6/8 marks awarded The answer here has explained the threat that commercial logging can bring to the different aspects of the tropical rainforest ecosystem. There is a focus on the impact that logging could have on the area and the deforestation that might result. However, the answer could have gone into a bit more specific detail and could have registered more facts and figures of this impact. More depth in relation to what is happening in the Monteverde Cloud Forest could have been listed. There is some good detail about the trophic structure and this is what allows the answer to move into Level 3.

(c) **With reference to a regional case study of a tropical rainforest ecosystem, describe its zonal soil and the processes of nutrient cycling.** (18 marks)

ⓔ Most of the 'big' 18 mark essay-based questions that appear at the end of each question require students to go into detailed reference of a case study. In this case, the question requires knowledge of the regional case study in tropical rainforests. This will likely be based on the Amazon basin. The answer must deal with both zonal soils and nutrient cycling.

Level 3 (13–18 marks): Answer will refer to an appropriate case study with accurate reference to the zonal soil (oxisols) and to the main components of nutrient cycling in the rainforest. There should be a detailed discussion about each of the layers within the soil, as well as all the key stores and flows in the nutrient cycle. Answers should use a high level of appropriate terms.

Level 2 (7–12 marks): All the main features of the answer will be present but the depth of the description and/or the links to the specific case study will not be present.

Level 1 (1–6 marks): An answer that makes very little or no use of the case study material would be confined to this level. An answer that lacks any mention of either the zonal soils or the nutrient cycling will be confined to this level.

(c) The Amazon basin is an area in South America that is covered in huge amounts of tropical rainforest. It covers over 7 million square kilometres and is one of the more lush and diverse areas on the planet. There are an estimated 390 billion trees and over 16,000 different species of animal. This makes the ecosystem very diverse and complicated.

The tropical rainforest of the Amazon has developed a particular type of zonal soil. A zonal soil is a soil where the soil profile has been shaped by the local climate and vegetation. In this case the hot and wet climate of the area has made the soil what it is today. The main type of soil in the Amazon is an oxisol. These soils have formed over a very long time and can be deep — up to 20 m. The top of the soil will contain a thin layer of humus. Although there can be a lot of leaf litter in the forest, the leaves will decay and decompose very quickly and few of the nutrients will make their way back into the soil. The top of the B horizon is often a red clay which is full of iron and aluminium. Deeper down into the soil the lower B horizon will start to lose its colour as the nutrients and salts in the soil are flushed through the soil due to leaching (because of the heavy rain). The C horizon can be deep where the parent material will often be weathered down. The soil will often become very acidic due to the leaching, which again makes it very difficult for vegetation to grow.

The nutrient cycle in the Amazon rainforest is often dominated by the excessive amount of biomass in the ecosystem. Plants, trees and vegetation are the main feature of the ecosystem. The leaves will fall to some extent and create some leaf litter. Some will decay and add nutrients to the soil but some will also be removed as part of the river run-off. The soil will not retain much of the nutrients. There is some weathering and some leaching in the system but most of the nutrients are lost to the system and this means that this ecosystem is fragile and will break down quickly.

ⓔ **12/18 marks awarded** This answer is at the top of Level 2. This answer contains some good reference to the regional case study. There is some good detail evident in the first paragraph that helps set the scene for the rest of the answer, but a better answer would have more specific details. The answer about soils in the Amazon is very good and is filled with specific detail, but the answer about nutrient cycling is not as strong and would require more specific depth to make sure that this deals with the stores/inputs/transfers and outputs that make up the system.

■ Option C

Question 3 Dynamic coastal environments

(a) Study the photograph which shows an example of hard engineering. Describe how this type of hard engineering works to protect the human and physical environment.

(9 marks)

ⓔ Most answers will note that this is a sea wall and they will give some sort of answer that describes how the wall works. Answers need to go into depth. There will be a temptation to talk about the sustainability of the engineering strategy, but the question clearly requires the candidate to describe how the sea wall can protect the human and the physical environment. This means that they might use some examples from a case study to back up what they are saying.

Level 3 (7–9 marks): The answer makes accurate reference to both the human and the physical environments that will get some protection from the installation of a sea wall.

Level 2 (4–6 marks): The answer will maybe have a detailed explanation about one of the two environments but the second will be less detailed. More detail or reference to particular places where this is in operation might be required.

Level 1 (1–3 marks): Answers here will show a limited understanding of the concepts of how the sea wall can protect the coast. There might be only a basic description of how this works and reference will only be made to one of the environments.

Student answer

(a) The picture shows an example of a sea wall. This is one of the most popular hard engineering strategies and is used around much of the coast of Northern Ireland. For example, in Portrush these measures have been used to protect the human and physical environments. The sea wall here will be used to absorb and reflect the wave energy and this will stop the coast from being eroded away. This was an expensive measure when it was first built and there have been additional costs to keep maintaining the wall to the highest possible standards. The beach can be eroded away as well, which can cause problems for the local council. Tourism is an important industry in Portrush and although the sea wall creates a nice wall for people to be able to walk along, the big problem is that often the sea wall will cause the sand from the beach to be removed further along the coast and people will have nowhere to sit.

ⓔ **5/9 marks awarded** This is a mid-Level 2 answer. There is some appropriate detail about how the sea wall works but there is really only explanation in relation to the human impact on the environment. The student has not gone into the reasons why the wall might be used to try to protect the physical environment and how it might protect fragile environments or even stop erosion at places where the bedrock might be soft and more liable for fast rates of erosion.

(b) Explain how any two of the following landforms are formed:

▪ **headlands**   ▪ **cliffs**   ▪ **beaches**   ▪ **spits**   (8 marks)

ⓔ Candidates need to present a full explanation of any two of the landforms from the list in the question. Each answer should contain an appropriate amount of detail with terminology and specific explanation of how the landform has been created. There should be discussion of the specific types of erosion that have been at work in order to leave the particular formation.

Level 3 (6–8 marks): Two of the landforms have been identified and explained in good depth with accuracy and a clear explanation of their formation. Specific details and information about how erosion has shaped the coast are present. There is a good use of terminology throughout.

Level 2 (3–5 marks): Two of the landforms have been identified and there is some explanation of their formation. Depth and/or the amount of details might be restricted.

Level 1 (1–2 marks): The answer here might lack an explanation of the two landforms or the amount of depth in the explanation for the two might be weak.

(b) Cliffs are one of the most obvious features along a coastline. Cliffs are slopes that are found at places along the coastline where wave erosion plays an important part. The waves will smash onto the rocks in the cliff and they will form a wave-cut notch. Over time, as the waves continue to smash into the cliff this notch will grow bigger and undercutting will take place. As the material becomes unsupported it will eventually collapse and this will leave a straighter and steeper section of shore exposed to the sea. All types of erosion will work on the cliff face at one time or another (with attrition and hydraulic action being the main ones) and the rock face will experience further action due to weathering from the wind and freeze-thaw. As the cliff face moves backwards a wave-cut platform will be left behind at the low tide mark.

Headlands are fingers of land that are made up of harder, more resistant rock. They will jut out from the coastline at a 90° angle from the sea. They are caused when there are two types of rock in an area. The softer rock will be eroded at a faster rate than the harder rock. The soft rock will be eroded away to form a bay and the headland will be left.

ⓔ 5/8 marks awarded There is a lack of balance in the answer. The detail in the answer about cliffs is good and contains the appropriate amount of detail that is required for an answer like this. However, the answer for headlands, although it contains some detail, needs a bit more depth to enable the answer to go into Level 3. There needs to be more balance between the two answers.

(c) **With reference to your LEDC regional study of a coastline under threat from sea level rise, describe how climate change could impact the human and physical environment.** (18 marks)

ⓔ The question demands that students describe an LEDC case study that is under threat due to climate change and sea level rise. The answer should develop in such a way that there is a detailed description of the different ways that climate change will *impact* both the human and physical environment.

Level 3 (13–18 marks): Answer will refer to an appropriate case study with accurate reference to each element of the question. There will be a full description and explanation of the impacts brought to the physical and human environments at the coast due to climate change. Answers should use a high level of appropriate terms.

Level 2 (7–12 marks): All the main features of the answer will be present but the depth of the description and/or the links to the specific case study will not be present. There might be some imbalance in the response to both the physical and human environments; more detail and precise case study facts might be required.

Level 1 (1–6 marks): An answer that makes very little or no use of the case study material would be confined to this level. If one of the elements of discussion of the impact on human or physical environments is missing, the response will be limited to Level 1.

(c) Over the last 20 years the impact of sea level rise on coastlines around the world has become increasingly obvious. Sea levels have been rising globally by around 2 mm each year.

Kiribati is a small island nation found in the Pacific Ocean. The population of over 100,000 people is spread over 33 low-lying atolls and reef islands. Sea levels in Kiribati have been rising at a higher level than the global average — at around 3.7 mm since 1992.

Impact on the physical environment

The islands of Kiribati are all very low-lying atolls which are ring-shaped coral reefs surrounding a lagoon. These islands are very low-lying so any slight change in sea level can have a huge impact. In recent years there has been an increase in the amount of coastal flooding. In 1997 Kiritimati island was devastated by an El Niño event that brought heavy rain and flooding across the area, causing a 50 cm rise in sea levels. Over 40% of the island's coral reef died and 14 million birds left the island. Another big tide in 2005 brought a storm surge 2.87 m high which wiped out villages and put pressure on fresh water supplies. This was another big impact that might occur again in the future. Kiribati has got a very sensitive supply of fresh water and when the sea water flows on the land it can contaminate the sources of fresh water and make life on the island more difficult. Some islands in the Kiribati range have disappeared due to flooding: Tebua Tarawa and Abanuea.

Impact on the human environment

The people who live on the islands are scared that in the near future there will be a really big sea flood that will wipe out any way for them to stay living on the island. There are big threats that over 50% of the population will not be able to survive and the country's only airstrip will disappear due to erosion. Some of the causeway roads that link some of the islands are flooded from time to time. The local government does not have enough money to spend protecting the land. Some of the local food stuffs are under serious threat as well – in order to protect food supply the government even took the step of buying land on one of the Fuji islands so that they always had food. As some of the water started to get contaminated by the seawater in the 1970s – some of the small villages in Kiribati have been moved and some have been moved further inland so that they are not under threat from the floods. These pressures have all made life in the islands less inviting and less sustainable than they have been in the past and some of the local people have moved away from the islands.

ⓔ 13/18 marks awarded This answer includes some useful points and argues clearly how the physical and human environment might be impacted in Kiribati. The answer gets into a low Level 3. There is a good balance between the two elements of the question and the candidate has got an obvious understanding of how the people and the environment of Kiribati are being affected by climate change and sea level rise. There are at least two points argued in both sections. The answer could have gone into a bit more depth and length which would have further increased the score.

■ Option D

Question 4 Climate change: past and present

(a) Study the map on the next page which shows the drumlin field in Ribbledale, northern England.

Describe the distribution of the drumlins formed as the glacier retreated and explain their orientation and nature.

(9 marks)

ⓔ Level 3 (7–9 marks): The answer clearly identifies the distribution of the drumlins and explains in detail both their orientation and nature.

Level 2 (4–6 marks): Either the descriptions or explanations are lacking in depth and detail, or one aspect of the question is not addressed fully.

Level 1 (1–3 marks): One key element is missing entirely. Alternatively, there is very limited use of the resources.

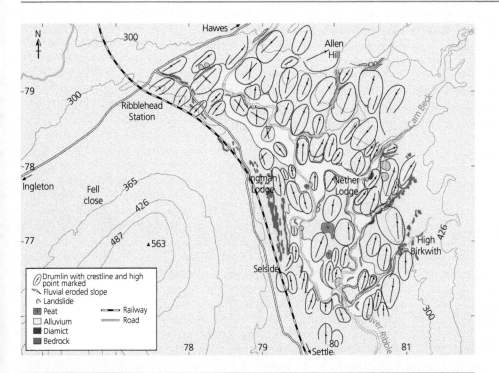

Student answer

(a) The drumlins are all located in the River Ribble valley to the east of the map. They also are mostly on the lower ground. To the west, almost all of them are below the 300 m contour. Those that go above it to the northwest of the map appear to have been cut off by the railway line. To the east of the map, a few of the drumlins are found just above the 300 m contour line, but none are above the 420 m contour line. This clearly shows that these are lowland glacial features.

The orientation of the drumlins runs initially from northeast to southwest, before turning to a more north to south orientation as the River Ribble abruptly changes direction. Their steeper stoss end is found to the north, and the more gentle lee end to the south. This seems to indicate that the direction of ice flow was following the River Ribble valley from the north of the map towards the south. As the glacier advanced, subglacial sediment was deposited forming the drumlins. However, the ice moving above these shaped the islands of deposits. It smoothed them off in the direction of the ice flow (hence their general north–south orientation) and it caused the stoss end to be shaped into a steeper slope, while allowing the lee end to taper off more gently.

e **9/9 marks awarded** All aspects of the question are addressed fully. There is full and detailed use made of the resource. The answer is clearly structured and there is good use of appropriate terminology.

(b) With reference to places for illustration, discuss the evidence for past climate change. (8 marks)

ⓔ **Level 3 (6–8 marks)**: At least two sources of evidence are discussed in detail and the links to climate change are explicit. Reference to places included.

Level 2 (3–5 marks): At least two sources of evidence are discussed, but neither is developed in detail, or the answer is imbalanced, or there is no explicit link to climate change. Alternatively, a more detailed answer that lacks references to places will be here.

Level 1 (1–2 marks): Only one source of evidence is offered.

> **(b)** One source of evidence is from ice cores. These are dug down into the ice and the air in the ice is looked at. Scientists can work out the temperature from these bubbles of air. Samples from Greenland date back 250,000 years, samples from Antarctic go back even further to around 800,000 years.
>
> Also, ocean-floor deposits, for example the Atlantic Ocean. These are drilled from the bottom of the sea and scientists can look at them and figure out from the fossils in them just what the climate was like. They go back much further than ice cores to over 2.5 million years ago.

ⓔ **4/8 marks awarded** Two pieces of evidence are examined, and places referred to. But the answer lacks depth and detail, and there is no link made to the climate change.

(c) Distinguish between mitigation and adaptation in tackling current climate change and discuss examples of mitigation and adaptation strategies that are being adopted, making reference to places for illustration. (18 marks)

ⓔ **Level 3 (13–18 marks)**: Both aspects are addressed with depth and detail and full understanding shown. Reference to places included.

Level 2 (7–12 marks): Both aspects are addressed but depth and detail are more limited, or the answer is imbalanced. References to places will be limited.

Level 1 (1–6 marks): If any one of the main elements is missing (adaptation, mitigation, reference to places), the answer will be confined to this level.

> **(c)** Mitigation uses strategies including carbon capture and reducing greenhouse gas emissions to tackle the causes of climate change. In contrast, adaptation seeks to take actions to reduce people's vulnerability to the impacts of climate change.
>
> One mitigation strategy is carbon capture. It tries to collect carbon dioxide from the places where it is being emitted, including power stations, and then takes it away somewhere else where it can be sorted. These include

underground stores. The carbon dioxide is dissolved in water and reacts with the underground rocks to form carbonates. This is currently being explored in Iceland. In 2016, a new approach to putting the carbon dioxide underground resulted in carbonates being formed in only two years, rather than the centuries previously expected. As a result of this good news, the Icelandic government now plans to bury 10,000 tonnes of carbon dioxide per year.

There are concerns about how effective this strategy will be, however. Some fear that the carbon dioxide will escape back into the atmosphere in the form of methane. In addition, there is energy needed to capture carbon in the first place. For example, for coal power stations, between 24% and 40% extra energy is needed. This increases the amount of fuel needed to produce the same amount of power.

Another mitigation strategy is to reduce greenhouse gas emissions. For example, you can seek to cut back the amount of carbon dioxide going into the atmosphere by using renewable sources such as wind, HEP, solar and wave energy. Also, you can try to make energy production more efficient by reducing dependence on coal (a high polluter). In China, traditionally much of its energy came from coal. However, it has been trying to reduce dependence on coal. In 2015, coal production fell by 3% and low-carbon energy production in the country has been increasing by more than 20%.

You can seek to reduce greenhouse gas emissions by reducing the burning of tropical rainforests. The REDD initiative is one attempt to do this. For example, Argentina lost around 15% of its forested area. The country became a REDD partner in 2009 and has been working to cut back on forest loss. This reduces the amount of carbon dioxide going into the atmosphere.

Secondly, adaptation strategies. Adaptation can take different forms. One is to introduce more resilient farming practices. Due to climate change, there is a risk of drought in parts of Africa. So, in Niger, small basins have been used to collect rainwater and this has led to a three- to four-fold increase in crop yield. A warmer world will have more energetic storms, and the UK is investing in flood and coastal defences to help prepare for increased flood risks.

Sea level rise is another major threat. In Bangladesh, they have been using strategies such as coastal embankments and salt-water resistant crops which are more resilient to sea water flooding. Around 1 million at risk people in Bangladesh have already been helped.

e **12/18 marks awarded** There are many strong elements to this answer: it clearly tackles all the major elements of the question, and the references to places are very good and integrated throughout. However, the answer is not as well balanced between mitigation and adaptation as it might be, and so it drops to top Level 2.

Knowledge check answers

1 Both the lithosphere and asthenosphere are made of solid rock. The key difference is that the asthenosphere will flow, and thus the convection currents triggered by heat from the mantle can rise through them, whereas the lithosphere is brittle. This means that, when it is subjected to forces of movement associated with tectonics, at first it will bend, but ultimately it will break. This is why earthquakes occur in the lithosphere.

2 The heat being transferred upwards via the convection current heats the lithosphere above it. This makes it less dense and more buoyant, so it rises isostatically to form the ridge.

3 To create volcanoes, magma is needed. This can be created in three ways: (1) decompression melting of the asthenosphere rock at a constructive margin as it rises with the convection current and experiences a fall in pressure; (2) subduction melting as the subducting lithosphere descends into the hotter asthenosphere at a destructive margin, aided by (3) hydration melting due to the sea water brought down by the process of subduction.

4 The youngest rock is found at the southeastern end of the island chain, corresponding to the active volcanoes now above the hot spot mantle plume. As you move further northwest, the age of the islands becomes progressively greater.

5 The first biggest hazard to human life comes from pyroclastic flows. Not only are they extremely deadly if you are caught up in one, but their speed means that, if you have not evacuated to a safe distance before they happen, you have very little chance of escape once one begins. Secondly, lahars can be extremely deadly, for similar reasons. The depth and debris they carry are very dangerous — but their speed means you need to respond very quickly should one be coming your way.

6 If the main impact of an earthquake is seismic shaking, and if the buildings are not made life safe, then this statement can be very true. In the Haiti earthquake in 2010, for example, over 250,000 people were killed, largely as a result of the collapse of non-life-safe buildings. However, the Japan earthquake in 2011 shows that significant loss of life can occur even in countries with very strict building codes. The main issue here was the devastating tsunami, coupled with problems with the Japanese authorities underestimating the size of the threat before the earthquake, and initially underestimating the magnitude of the quake when it occurred.

7 Tropical rainforest is found between 10° north and 10° south of the equator. Tropical grassland is found between 5° and 20° north and south of the equator and desert is usually found between 15° and 30° north and south of the equator.

8 Tropical rainforest annual average rainfall is around 2,000 mm, whereas the rainfall in the tropical grassland is around 830 mm and in the desert it is below 250 mm.

9 Heat is transferred to the air above and the air rises and cools. Rising air means that low pressure is located here, in the region known as the inter-tropical convergence zone (ITCZ). The rising air diverges and flows towards the poles (both north and south) and sinks again around 30° north and 30° south of the equator. This warms and produces high pressure.

10 Sustainable development is described as being socioeconomic progress that meets the needs of today's population without harming the ability of future generations to meet their needs.

11 Waterlogging is when there is too much water on the land. The soil becomes saturated with moisture and this means that the soil cannot get the oxygen that it needs for plant growth. Conversely, salinisation occurs when the salts and minerals are pulled to the surface through the process of evaporation and capillary action. This makes the soil useless and crop harvests will be reduced.

12 Traditional farming practices by native people using slash-and-burn techniques are more sustainable than large-scale logging and industrial clearance. Usually natives will only clear a small area, burn the vegetation to release the nutrients back into the soil and will then only farm the area for a few years before moving on and allowing the small areas to be reclaimed by the forest.

13 Ecotourism is responsible travel to natural areas that conserves the environment and improves the well-being of local people by creating an international network of individuals, institutions and the tourism industry and by educating tourists and tourism professionals

14 In deep water, a wave transfers energy only. However, as it enters shallow water, the lower parts of its rotation come into contact with the sea bed and begin to experience friction. This slows this wave crest down, causing the wave to compress, thus decreasing its length while at the same time increasing its height. Eventually, the top of the wave becomes too steep and it breaks. At this point, the wave carries matter as well as energy. This includes sediment along the shore, and so breaking waves are vital in the process of coastal transportation.

15 Constructive waves produce spilling breakers, whereas destructive waves produce plunging breakers. Constructive waves have a strong swash, whereas destructive waves have a strong backwash. Constructive waves tend to carry material up the beach, increasing gradient, whereas destructive waves carry it down the beach, decreasing gradient.

16 A spit is a linear ridge of deposited material that builds out from a coastline where there is a sudden change in the orientation of the shore. For example, spits can occur as they build out across the mouth of an estuary. They are only attached to the coast at

one end. On the other hand, where a spit builds out across the entire estuary, joining on to the coast at the other side, then a bar is formed.

17 Eustatic changes in sea level are global in scale and follow major worldwide events such as the end of a glacial period, when melting ice from the land enters the sea, raising global levels. In contrast, isostatic changes are local or regional in scale. They too may be associated with glaciation — during interglacials, with the weight of glacial ice having been removed from the land, it can spring back upwards. However, there are other more localised causes of isostatic change such as during an earthquake.

18 At the end of a period of glaciation, eustatic and isostatic changes can be occurring at the same time. For example, melting of land-based ice causes eustatic rises in sea level. At the same time, however, the land may be rebounding upwards isostatically now that the ice is melting, causing an isostatic fall. So it is the relationship between the rates of eustatic and isostatic change that determines what happens at a particular coastline. If the rate of eustatic rise is faster than the isostatic fall, then there will be a net rise in sea levels. This occurred in the British Isles in the first 6,000 years following the end of the last glacial period. For the past 4,000 years, however, in the northern half of the islands, the land is continuing to rise upwards, causing an isostatic fall in sea levels there.

19 With Shoreline Management Plans, there is a strong emphasis on sustainability into the twenty-second century. As a result, hard engineering will only be used where there is no other choice (for example, where high value urban land needs it for protection) as hard engineering tends not only to require ongoing maintenance investments, but also can interrupt the natural coastal processes. If, for example, a section of a sediment cell is not acting as a major source of sediment for elsewhere, and if there is high value urban land behind it, then hard engineering may be chosen. However, soft engineering is the preference, as it is not only cheaper in the long term, but it will allow the whole sediment cell to operate in a much more natural manner.

20 For most of the history of the Earth, the climate has been warmer than it is now. This meant that there were no permanent ice caps on the poles. However, before our current ice age, evidence exists for five

ice age periods (defined as times when the poles were covered in permanent ice), including one which seems to have covered the entire planet around 800 million BP. About 50 million years ago, the climate changed from warmer and more stable towards cooler and more erratic. The last major glacial period is known as the Pleistocene (from 2.6 million to 11,700 years BP). More recently, we are in the interglacial known as the Holocene.

21 Ice core records do not go back as far in time (up to 800,000 BP) as ocean-floor deposits (2.6 million years BP — or the entire Pleistocene). However, ice cores give more detailed year-by-year records than ocean-floor sediments do, as they are formed on an annual basis.

22 First, subsequent snowfalls put weight and pressure on the snow, increasing its density to form firn. When the snow first falls, it traps air in the spaces within it. As some snowmelt occurs, however, this water percolates into these spaces and subsequently freezes. This increases its density, gradually turning it into ice. This can take decades in the Alps, and centuries at the poles.

23 Glacial till is unsorted and more angular. This is because it has all been transported by the massive glacier which is able to transport sediment of a wide range of sizes. In contrast, fluvioglacial deposits are sorted and more rounded, as they have been transported by meltwater streams found under the glacier.

24 Terminal moraines mark the furthest extent of the glacier and run parallel to its leading edge. As the glacier retreats, it may stop at a location for a while, forming another 'terminal-like' moraine there — in this case, however, this is known as a recessional moraine. Due to climatic variability, the glacier having retreated a bit may then move forward again. If it does so, it can scour forward the old terminal moraine, compressing it upwards to form the steeper and taller push moraine.

25 Mitigation attempts to reduce further emissions of atmospheric pollutants that contribute to climate change, whereas adaptation seeks to implement policies to cope with the impacts of climate change. Most agree that both approaches are needed: even if we are successful in our mitigation strategies, there will still be impacts of the climate change that has already happened for us to address, and thus adaptation is vital too.

Index

A

abyssal plain   8, 11
adaptation to climate change
    87–88
air circulation, Hadley cell   34–35
air pollution and climate
    change   84–85
Amazon basin, case study   43–46
arches   55, 56
Arctic sea ice, loss of   84
ash clouds, volcanoes   17–18
asthenosphere   6

B

bars   59
beaches   56–57
    drift-aligned   54
    nourishment   67
    raised   62
    swash-aligned   53
    wave impact   51–52
Benioff Zone   13, 14, 23, 28
biomass
    and nutrient cycling   41, 44
    tropical biomes   31–34
biomes   28–29
breaking waves   51–52
building codes, Japan   23, 26–27

C

carbon capture   86–87
carbon dioxide
    attempts to reduce emissions   90
    rising levels of   84–85
    volcanic emissions   18, 74
case studies
    Amazon basin, Brazil   43–46
    earthquake, Tohoku, Japan   26–28
    irrigation along Indus river,
        Pakistan   39–40
    lowland glacial region in south-
        east NI   81–83
    Monteverde Cloud Forest, Costa
        Rica   47–49
    north Norfolk coastline   68–70

sea level rise in Kiribati   63–64
volcanic eruption, Mt
    Pinatubo   20–21
caves   55–56, 62
CBA (cost–benefit analysis)   65
central rift valley   10, 11
cliffs   54–55, 62
climate change
    causes of   73–75
    current change, human
        causes   83–85
    evidence for   72–73
    glacial and fluvioglacial
        landforms   79–83
    Holocene period   75–76
    international action on   88–91
    lowland glacial landscapes   77–79
    managing   86–88
    natural processes   71–72
    present and potential
        impacts   85–86
    questions and answers   104–107
climate, tropical biomes   30–31
coastal environments
    management and
        sustainability   64–71
    processes and features   49–60
    questions and answers   100–104
    regional coastlines   60–64
collision margins   12
composite volcanoes   15
conservative margins   11–12, 23
constructive (divergent)
    margins   10–11, 15
constructive waves   51–52
continental drift   75
convection currents   10, 13
cost–benefit analysis (CBA)   65

D

debris   78
decompression melting   15, 16
deep ocean trenches   8, 13
deep sea trenches   14
deforestation   47, 48, 87

deposition   56, 79, 80, 81
desert   29–30
    biomass   31, 32, 34
    climate   31
destructive margins   9, 13–14
    volcanic activity at   15
destructive waves   51–52
dilation theory   25
drift   79
    continental drift   75
    longshore drift   54, 58
drift-aligned beaches   54
drip irrigation   37, 38, 40
drumlins   80, 81, 82
dunes   57–58
    regeneration   67
dykes   10, 11

E

earthquakes   22–23
    global pattern of   9
    impacts of   23–24
    prediction attempts   24–25
    preparation for   25–28
Earth's structure   6
ecosystem, defined   28
    ecotourism   47, 48–49
emergent landforms   62
Environmental Impact Assessment
    (EIA)   65
epicentre, earthquakes   22
erosion
    coastal   53, 55–56, 68–70
    fluvioglacial   78
    glacial   77–78
erratics   80
eskers   81
eustatic sea level changes   60

F

fetch   50
fjords   61
floods/flooding   40, 61, 63
    adaptation and prevention   88

fluvioglacial landforms   79, 81
fluvioglacial processes   77–79
focus, earthquakes   22
fold mountains   12, 14
Fukushima nuclear meltdown   27

G
gabions   67
geothermal energy   19
Gersmehl model of nutrient cycling   42, 44
glacial landforms   79–80
glacial processes   77–79
glacier retreat   84
global warming   83–85
grassland *see* tropical grassland
gravity irrigation   37
greenhouse gas emissions   84–85
   attempts at reducing   86–87
groynes   67, 69–70

H
Hadley cell   34–35
hard engineering   66–67
Hawaiian Islands, linear pattern   16
hazards map   20–21
hazards of volcanic activity   17–18
headlands   55
Holocene period   72, 75–76
hot desert   29–30
hot spots, volcanoes   16
hydration melting   13, 15

I
ice ages   71, 72, 74, 75
ice cores   72
ice sheets
   formation of   77
   loss of   84
   and sea level changes   60, 61
Intergovernmental Panel on Climate Change (IPCC)   83–84, 90–91
irrigation   37

environmental and socioeconomic impacts   37–38
   Indus river, Pakistan   39–40
   solutions to problems   39
island arcs   13, 14
isostatic sea level changes   60–61
ITCZ (inter-tropical convergence zone)   34, 35–36

J
jökulhlaup   18

K
Kiribati Islands, sea level rise   63–64
Kyoto Protocol   88–90

L
lahars   17, 21
large-scale development
   Amazon rainforest   43–46
   negatives of   45–46
   positives of   46
   threats   41–42
lava flows   15, 17
LEDCs   85, 87, 88, 89
life safe buildings   23, 25
linear pattern of islands   16
liquefaction   23–24
lithosphere   6
Little Ice Age   72, 74, 76
longshore drift   54, 58
lowland glacial landscapes   77–83
L waves, earthquakes   22–23

M
magma   9, 15, 16
managed retreat   65, 67, 70
mantle plume   16
MEDCs   85, 88, 89, 90
mid-ocean ridges   8, 10
Milankovitch cycle   73–74
mineral creation   18
mitigation of climate change   86–87
Monteverde Cloud Forest, Costa Rica   47–49

moraines   79, 80
mountain building (orogeny)   61, 75
Mt Pinatubo eruption, Philippines   20–21

N
Norfolk coastline erosion   68–70
notches   55
nutrient cycling   41–42, 44

O
ocean floor age   8
ocean-floor deposits   73
ocean floor relief   8
ocean ridges   8, 11
ocean warming   84
orogeny (mountain building)   61, 75
outwash plains   79, 81

P
palaeomagnetism   8–9
Pangea supercontinent   6–7, 75
plate tectonics
   processes at plate margins   10–15
   questions and answers   94–96
   seismic activity   22–28
   theory of   6–10
   volcanic activity   15–22
poisonous gases in magma   18
pollen analysis   73
primary productivity   31
P waves, earthquakes   22–23
pyroclastic flows   17

R
rainfall
   Hadley cell and ITCS   34–35
   tropical biomes   30, 31, 32, 33
rainforest *see* tropical rainforest
raised beaches   62
relict landforms   62
resilience   88
revetments   66, 70
rias   61
rip-rap   66

Index

S

salt marshes 58
savanna *see* tropical grassland
sea-floor spreading 10, 11
sea level changes 60–61
sea level rises 61–62
  adapting to 88
  due to climate change 62
  Kiribati case study 63–64
  north Norfolk case study 68–70
sea walls 27–28, 66, 69–70
seismic events 22–23
  prediction of 24–25
  preparation and response to
      25–28
seismic gap theory 24–25
seismic shaking 23
shield volcanoes 15
Shoreline Management Plans
    (SMPs) 65, 66, 68
socioeconomic benefits of volcanic
    activity 18–19
socioeconomic hazards, volcanic
    activity 17, 18
soft engineering 67–68
solar energy/output 74, 84
spits 58–59
spray irrigation 37, 38
stacks 55, 56, 59, 62
stumps 55, 56
subduction 10, 12, 13
subduction melting 13, 14, 15

submergent landforms 61
sunspots 74, 83
sustainable development 37, 47
swash-aligned beaches 53
S waves, earthquakes 22–23
swell waves 50

T

temperature changes *see* climate
    change
temperatures, tropical biomes 30–31
tephra 17
till 79
Tohoku earthquake, Japan
    (2011) 26–28
tombolos 59
tourism 19, 68
  ecotourism 47, 48–49
trade winds 34, 35
transform faults 11–12
transportation
  coastal 53, 54
  glacial and fluvioglacial 78
trophic structure, tropical
    forests 41, 43–44
tropical biomes 28
  biomass 31–34
  climate of 30–31
  distribution 28–30
tropical ecosystems
  arid/semi-arid 37–41
  forest environment 41–49
  locations and climates 28–36

  questions and answers 97–100
tropical grassland 29
  biomass 31, 32, 33–34
  climate 30, 35–36
tropical rainforest 29
  biomass 31, 32–33
  climate 30, 36
  large-scale development 41–46
  sustainable development 47–49
  zonal soil 42
tsunamis 24
tsunami sea walls, Japan 27–28

V

volcanic activity 15–16
  and climate change 74
  hazards and benefits 17–19
  preparation and response to
      19–21
volcanic ash 17–18
volcanoes, formation of 14
vulnerability, reducing 87–88

W

wave action 49–52
wave erosion 53
wave refraction 52–53
waves, seismic 22–23
wave transport 53
wind waves 50

Z

zonal soils 42, 44–45